Biohacking

How to Boost Neurogenesis and Rewire Your Brain

(Successfully Employing Biohacks to Improve Your Health Life & Wellbeing)

Angelo Finger

Published By **Bella Frost**

Angelo Finger

All Rights Reserved

Biohacking: How to Boost Neurogenesis and Rewire Your Brain (Successfully Employing Biohacks to Improve Your Health Life & Wellbeing)

ISBN   978-1-998038-33-6

No part of this guidebook shall be reproduced in any form without permission in writing from the publisher except in the case of brief quotations embodied in critical articles or reviews.

Legal & Disclaimer

The information contained in this book is not designed to replace or take the place of any form of medicine or professional medical advice. The information in this book has been provided for educational & entertainment purposes only.

The information contained in this book has been compiled from sources deemed reliable, and it is accurate to the best of the Author's knowledge; however, the Author cannot guarantee its accuracy and validity and cannot be held liable for any errors or omissions. Changes are periodically made to this book. You must consult your doctor or get professional medical advice before using any of the suggested remedies, techniques, or information in this book.

Table Of Contents

Chapter 1: Sleep Tracking

Sleep is a important factor for each mental and bodily nicely-being. Without good enough sleep, you can't artwork successfully every day. Not honestly does a lack of sleep have an impact on your overall performance at paintings and domestic, it can, however, even bring about you some excessive clinical troubles. Poor first-class sleep can cause mental health troubles like tension and depression, now not to say bodily issues like excessive blood strain and cardiovascular sickness, which can have enduring and important consequences.

Nevertheless, while we apprehend that we need to all be getting as a minimum eight hours of sleep every night time time, it may be difficult to accomplish that intention. Whether you figure shifts, have circle of relatives duties, or are having a difficult time to get all your university or college paintings finished, becoming in enough relaxation may be a excessive impediment. So, how are you going to restore that hassle? The response might be monitoring your sleep styles.

Thanks to the maximum modern-day generation, you can have get right of access to to sleep tracking functionality inside the residence. You can now buy wearable sleep trackers which might be going to music your sleep styles to make sure that you may end up being more familiar with the length of time you sleep, the ranges of sleep that you get to, and the satisfactory of sleep.

Nowadays, we will tailor more matters in our lifestyles than within the beyond. We can customise our garb, telephones, and homes, it makes feel so that you can individualize your sleep patterns too. Not certainly everyone requires the precise same amount of sleep each night time, however, if you track your styles, you can come to a extra information of the quantity of hours this is proper for you.

If you're having sleep issues, a wearable sleep tracker goes to help you in identifying the motive in the back of your sleep issues. At one time, the quality preference was to visit a nap laboratory to get an professional assessment. Now, you may have similar capability on your snug mattress. With the provision and precision of cutting-edge sleep trackers, you could discover bothersome styles and modify your workouts for the higher.

In case you music your sleep styles, you'll additionally begin to wake at the advanced time. A lot of the top notch sleep trackers have a clever alarm to wake you at the identical time as you're within the direction of the lightest sleep phase. This prevents you from waking up dazed and irritable. Rather, you can feel rested and organized from the start for a greater green day.

Sleep Tracking Benefits

If you purchase a snooze tracker, you could enhance your sleep wonderful and, therefore, the way of existence you can take satisfaction in. More people are nervous and harassed than ever earlier than, so getting good enough notable sleep is vital.

If you are sleep disadvantaged over extended durations, you are vulnerable to clinical troubles, each bodily and intellectual. Cardiovascular contamination, type II diabetes and breathing problems have definitely all been associated with snoozing problems.

While we're virtually acquainted with what we're doing ultimately of the day, at some point of the night time, our regimens often get forgotten. We're used to preserving music of behavior at a few stage in the day, from what we consume to truly how masses workout we get, so we must begin to do the same element at some stage in the nighttime hours.

If you track your sleep, the quality advantage is that you could start to grow to be aware about hyperlinks in among your sleep patterns and normal health. For instance, you'll discover whether or not or now not or now not eating coffee or eating caffeine adversely influences your sleep or whether or not or not or now not the alcohol that you consume affects the best of your relaxation.

Chapter 2: Blue Light

More humans are virtually carrying out being greater familiar with how blue mild can effect our our our bodies, however with prolonged device use, we are subjected to more of it than ever earlier than. To satisfaction in maximum applicable well-being, we need to discover strategies to reduce the quantity of blue moderate we allow ourselves to be subjected to. Why is that this the case? Keep studying and discover more.

Blue Light's Impact at the Body?

Comprehending how mild engages with our eyes is the trick to information why blue mild is so bad for our nicely being. Light is crafted from severa colored waves, which all have their severa energies. Red slight is on the begin of the plain moderate spectrum. This has low strength waves and is a notable deal an awful lot less hard on the eyes, especially all through the night.

Blue moderate, even though, has the finest strength waves and this makes it more difficult for the eyes to technique correctly.

While high electricity light waves are essential for our everyday lives, it can show to be unstable if we're subjected to them at incorrect times. High electricity light is acquired from the sun to manipulate sleep styles effectively. In the day, the moderate enters our eyes to discharge enzymes, deliver our stages of melatonin down and help us in awakening.

Melatonin manages our sleep cycles with the aid of manner of natural rhythms. Nevertheless, this cycle is all too easy to break. Extreme subjection to blue mild can interrupt your body clock cycle. This is because of the fact that it minimizes the melatonin degrees discharged via your frame. If your frame does no longer have sufficient melatonin at bedtime, you can not sleep efficiently and you emerge as being tired.

Blue mild is discharged with the aid of the use of monitors from laptops, pills and smartphones and in the end induces eye pressure, near-sightedness and dry, itching eyes. Even worse, blue slight affects the retina and the cellular anchor and can activate advanced macular degeneration at an earlier age. Some experts have genuinely even connected weight troubles to melatonin disturbance in addition to the improvement of some forms of maximum cancers. Discovering techniques to prevent excessive blue slight exposure is, for this reason, crucial.

What to Do About Blue Light?

Some makers of devices inside the imply time are acknowledging the damage that blue slight can set off and are starting to establish brand-new technological offerings to address the difficulty. Blue-filter covers are supplied for purchase for VR goggles, laptops and smartphones and a few gadgets have without a doubt now included "night time time time modes" into their model, which gets rid of the blue moderate at the same time as you make use of your system at night time time time hours to restriction your direct publicity.

Naturally, the apparent strategy to the blue moderate direct publicity problem is merely to stay clean of utilising any device in some unspecified time inside the future of the night time. Tablets and cell phones want to be stored out of the mattress room, and for numerous hours preceding to bed, we need to live some distance from any device usage. Sadly, this isn't continuously feasible or likely best. So, how can we avoid the difficulty?

The solution will be to shop for a set of blue-moderate protective glasses. These are created with an HEV clean out blanketed. These allow you to employ your devices every time you like with out a undertaking approximately direct publicity to blue slight.

Blue light blocker glasses appear to be a easy set of glasses, but, they have unique filters that save you excessive strength essential moderate from attending to the rear of the attention. They can each be provided as a standalone set of glasses or as a completely particular set of night time time-time glasses which might be used over a routine set of eyeglasses. In case you location the ones on your head round an hour preceding to going to sleep, they will be going to close out all of the blue slight released from your devices and LED lights supporting you to improve sleep every night time time.

Chapter 3: Drink a Different Kind of Water

Another beneficial hack that you could upload to decorate your manner of existence is to devour alkaline water. This is maximum of the present day discoveries within the nicely being and health vicinity, and it's miles one trouble which could speedy be blanketed on your existence.

All human beings understand that water is vital to each tissue, cellular and organ in our our our bodies. Nevertheless, some of us prevent operating to live as hydrated as we should with the useful resource of ingesting accurate sufficient glasses of water. Now, brand-new research has in reality found that we should consume greater water, however, the shape of water that we consume is further important. Alkaline water is concept to be the very amazing issue for us to devour.

Why is Alkaline Water So Good?

Water is made from oxygen and hydrogen, with the style of hydrogen ions inside the water being determined as a pH decide.

Alkaline water consists of lots less hydrogen ions and has a greater pH degree than fundamental faucet water. Water has a pH degree which varies in amongst 0 and 14. 7 is said to be unbiased, with an identical stability in among acidic and alkaline. When the water has a pH stage of beneath 7 it's far acidic, and greater than 7 makes it alkaline. Water from the faucet inside the United States typically has a pH of in amongst 4.Three and 5.3 relying upon in which you stay.

Those who proclaim the deserves of alkaline water nation that the larger shape of hydrogen ions assists in imparting greater hydration when contrasted to regular water, mainly when you have sincerely been exercise. They furthermore think that routine tap water that has an acidic pH triggers excessive acid to growth for your blood and cells primary to a chain of health problems.

They assume that alkaline water can reduce the quantity of acid on your bloodstream, presenting your metabolic method an increase, boosting your power tiers, reducing the growing older technique, enhancing your food digestion or even minimizing your bone loss. Some even u . S . A . That it can starve maximum cancers cells.

Alkaline water is an anti-oxidant that reduces the consequences of the complimentary radicals which cause cellular and DNA damage. Thanks to the tinier cluster period of the water, it is able to permeate the cells extra speedy to hydrate you better, and its greater popularity of alkaline minerals which includes magnesium, calcium and potassium assists to guarantee greater health.

Alkaline water is furthermore enough in oxygen which complements the amount of oxygen liquified within the blood, and thanks to its cleansing competencies it is able to do away with the accumulation of mucous on the partitions of the colon to enhance your frame's nutrient-absorbing talents.

This sort of water can get rid of the contaminants and acidic waste which has truly constructed up for your frame even as the adversely charged ions help in increasing your recognition, mental clearness, and strength stages. It may even assist in coping with your weight and in closing healthful because of the reality contaminants which can be usually found in faucet water are removed.

How to Use Alkaline Water Properly

If you're organized to experience the benefits that eating alkaline water can convey, you need to understand how you may gather it. While you should purchase bottled alkaline water in stores, you may create it your self with a water ionizer. A water ionizer is a compact tool that links to the deliver of water to your kitchen region. It plays low voltage electrolysis for your faucet water previous to eating it or the use of it on your kitchen for cooking or cleansing.

A water ionizer uses a specific accent that reroutes the water from the faucet through a pipe into the device. In the tool, water is filtered to eliminate the maximum conventional contaminants which might be determined in recurring tap water. Next, the water which has in reality been filtered hand down right into a chamber it's geared up with titanium electrodes included with platinum and it remains in this chamber in which the electrolysis takes place.

Positive ions (or cations) converge at the terrible electrodes. This develops decreased (or cathodic) water. On the alternative hand, the anions or negatively charged ions, converge on the wonderful electrode. This creates oxidized or anodic water.

The ionized water is guided to the tap on the identical time as the oxidized water is guided to each different pipe which ends up in the sink. The ionized water can then be applied for eating or cooking. As a gain, the oxidized water works as a disinfecting agent that is probably utilized for cleansing up utensils, palms and food. If you buy a water ionizer for your private home, you could satisfaction in all the benefits of alkaline water every day with out a hassle.

Chapter 4: Red Light Treatment

RLT or Red Light Therapy is a restorative approach which makes use of low-degree purple-moderate wavelengths to address loads of skin troubles like accidents, wrinkles and scars together with different situations. How can it assist us in enhancing our trendy fitness? Keep analyzing.

What Is Red Light Therapy?

Throughout the Nineties, Red Light Therapy end up used by researchers for cultivating plant life in area and it became in the course of this method that they determined the extreme moderate generated thru manner of purple LEDs (mild-emitting diodes) promotes photosynthesis and development of the plant cells. It emerge as then that pink moderate began out out to be researched for its viable blessings within the realm of drugs with tests being finished to see if

RLT boosted the strength in human cells to address bone density troubles, gradual recuperation of accidents and muscle atrophy.

There are severa sorts of RLT with each splendor and scientific applications. It can address extreme conditions like sluggish-to-heal injuries and psoriasis, along side splendor issues consisting of stretch marks and wrinkles.

Red Light Therapy skills thru generating biochemical effects in the cells to beautify their mitochondria. These are the cellular's powerhouse wherein the power of the cellular is produced. ATP is the energy-carrying particle that might be found within the cells and at the same time as RLT boosts the mitochondria's function, it creates greater ATP. Consequently, cells have greater strength which allows them to revitalize themselves, repair harm and feature higher.

Unlike unique IPL or laser remedies, RLT does not harm the floor location of the pores and skin. Rather, it promotes pores and pores and pores and skin regrowth to use a bunch of advantages such as:

- Promotion of harm restoration and tissue repair.

- Treatment of carpal tunnel syndrome

- Enhanced hair boom

- Decrease of psoriasis sores

- Stimulated recovery of accidents which might be gradual to get better

- Decrease of destructive results from most cancers remedies

- Relief of soreness and tightness in the ones experiencing rheumatoid arthritis

- Enhanced skin tone and boosted collagen for decreased wrinkles

- Fixing of solar harm

- Enhanced joint health in humans with osteoarthritis

- Prevention of repeating fever blisters

- Relief of ache and swelling

- Reduced scars

As you could see, there are some of motives why you need to consider having Red Light Treatment to decorate your regular health and health.

How to Take Advantage Of Red Light Treatment

You can find out purple moderate treatment in use for splendor capabilities in spas, tanning stores and gyms. Nevertheless, there are additionally some of FDA-authorized RLT devices that you can purchase for utilization in your private home. While the ones will now not be as effective as the ones which you could find out in scientific utilization, they artwork at preventing unwanted signs and symptoms of early growing older like excellent traces, age spots and wrinkles. If you choice to address a systematic scenario, however, you'll have to speak approximately the alternatives provided to you with a health practitioner to guarantee you experience final benefits.

RLT is pain-unfastened and stable if the devices are implemented efficaciously. There have actually been testimonies of some people getting burns because of machine rust, broken wires or due to the reality that they dropped off to sleep with their device nonetheless in place. Nevertheless, even as the recommendations are located effectively, pink moderate remedy is not risky. It is, regardless of the fact that, simply essential to utilize eye protection due to the fact pink slight can harm the eyes.

Considering that RLT has sincerely been validated to have appealing results even as it worries managing some of skin problems, it is nicely clearly certainly well worth at the side of red light treatment into your regular skin care software program. If you're worried approximately your signs, you would possibly need to are looking for recommendation out of your clinical professional to begin with to make sure that RLT is the very amazing choice for you, however for masses of people, the benefits of pink slight treatment make the investment in a pink moderate-emitting system properly well worth it.

Chapter 5: Be More Active

All human beings understand that we should get hundreds of sporting sports to stay in top bodily and mental shape. Nevertheless, but this normal facts, a number of us despite the fact that are not getting the encouraged quantity of pastime on an ordinary or weekly basis. The next crucial hack to embody into your existence is to encompass extra exercise every day. How are you in a position to perform this at the same time as you're on a decent time table? Here, we check some of the advantages of being more bodily and a few modern techniques to fir extra interest into your every day existence.

The Issues of an Inactive Way of existence

You'll most in all likelihood have heard inside the media that quite a few human beings are residing an inactive manner of life, but, what does this imply?

Inactive living is living in that you do no longer do enough exercise often. The present suggestions via the CDC are that absolutely everyone want to do no less than one hundred fifty mins of slight workout each day or, additionally, 75 mins of active workout. Walking 10,000 steps day by day is typically recommended to decorate your health and to restriction the feasible health threats which show up because of lack of exercise.

The WHO states that as a lot as eighty 5 percent of the area's populace isn't always sufficiently bodily lively and this makes the inactive manner of life the 4th foremost threat across the area. Typically, we're led to accept as proper with that eating healthily and taking some cardio exercising can balance out all the influences brought approximately via too much time taking a seat. Nevertheless, evidence now well-knownshows that during case you exercise session for half of of of an hour an afternoon, you continue to might not be capable of reduce the possible damage. The very exceptional choice appears to decrease the amount of time taking a seat and increase the amount of time we decide to shifting daily.

The inactive way of lifestyles leads to many poor impacts. Whether you are strolling at a desk or using a bus or taxi, you are placing your self on the threat of the following troubles:

- A more chance of setting up precise cancers

- A more possibility of struggling with specific cardiovascular issues

- A higher chance of putting in anxiety and depression

- A better opportunity of finishing up being obese or obese

- Greater immoderate blood stress

- Lowered skeletal muscular tissues

- Raised levels of cholesterol

It has definitely been approximated that, global, inactive manner of existence induces 7 percentage of all instances of coronary cardiovascular disease, 6 percent of all instances of type II diabetes, eleven percent of all times of breast most cancers, and 11 percent of all times of colon most cancers. It has honestly even been disclosed that a sedentary manner of existence leads to more deaths each 365 days than cigarette smoking.

We are extra inactive these days than we ever have surely been in advance than considering the truth that innovation has absolutely altered how we stay our lives. 50 years all over again, much less humans applied automobiles and had desk jobs. They moreover had greater physical entertainment sports activities and pastimes in region of viewing Television and gambling computer video video games. The amount of inactive obligations has truly grew through the use of manner of over eighty percent since the 50s, and even as we consist of into that the truth that we've longer common running weeks, that is a lot extra time devoted to being in a seated function.

It's obvious that coming across techniques to neutralize the detrimental impact of the inactive way of life is vital, but, thankfully, there are various biohacking modifications to beautify your health, health and fitness in well-known.

How To Be More Active

All of us understand that maintaining our physical fitness need to be a prime mission, however, we are likewise busier than ever earlier than in our lives. With responsibilities like looking after senior parents or children, a hectic social existence and annoying jobs, we're all below pressure in a frenzied tempo of lifestyles. Naturally, the maximum apparent method of getting extra active is to go to a fitness center or to reserve an hour every early morning or night time to exercising within the house. Nevertheless, this simply isn't always feasible for a few human beings.

Lots of human beings are nervous with the resource of the concept of heading to the fitness center, at the same time as discovering the time to suit bodily interest proper right into a normal recurring can be almost exceptional. For that reason, discovering techniques to become being more active on the identical time as placing about our routine sports sports is the very high-quality opportunity.

Here are multiple smooth hacks to exchange your every day regimen into a much greater wholesome one:

- Change to a status table in choice to a ordinary one. Office workers sense connected to their desks for an awful lot of the day, but, if you make the smooth amendment to status in desire to sitting, you could find out that you're tons less stiff and gradual whilst your day entails a end. Standing utilizes masses more muscle corporations whilst contrasted to sitting, and evidence has genuinely discovered that repute every thirty minutes and walking round can lessen your odds of passing away early. Even higher, it promotes lots better posture which, in flip, lowers anxiety and exhaustion while urging a super deal higher common overall performance and steadier respiratory. If you desire to take subjects a step similarly, why no longer alternate to a treadmill table as a substitute. This goes to can help you remain lots greater lively whilst you work, and you could jog or walk at some stage in your important paintings sports activities.

- Take the stairs in region of the elevator. Walking up a slope is an awful lot better for you than on foot on a flat ground area, so choose out the stairs for max advantage even as becoming active. Research well-known that during case you climb up the steps really 3 instances weekly, your cardiorespiratory bodily health goes to decorate. Your leg muscle tissue are going to come to be being extra powerful, and you could moreover burn more electricity for less tough frame weight protection.

- Include primary exercising routines into your routine. If you do now not have adequate hours inside the day to gain the gymnasium, embody a few muscle physical sports into your ordinary application as an opportunity. Doing squats at the same time as at your table or dips in your place of work chair will now not take a whole lot paintings and it's miles capable to help you in improving your ultra-modern bodily health and fitness. You can also even encompass small adjustments like stabilizing on one leg whilst brushing your enamel or doing custom designed push-u.S.A.Versus your counter top.

- Ditch the car. Rather than the usage of to paintings or the shop, attempt cycling or on foot as an alternative. You'll discover that it is going to carry you physical and mental advantages.

- Utilize a resistance ball in desire to a regular chair. Whether you are at art work or in your house, changing your fame chair for a resistance ball is going to help to right away align your spinal column, beautify your posture and encourage you to stretch and drift greater frequently. You can also even do some little wearing events concurrently like changed sit down down-americato contain middle muscle strength.

- Take brief strolls for the duration of the day. Throughout your lunch spoil, in place of vegetating at your table, take a short walk throughout the block alternatively. Simply a ten-minute stroll each day can provide you favorable bodily and mental benefits. An workout does now not want to take an hour. Simply 10 or 15 minutes of physical hobby gives advantages too and goes to get your coronary heart pumping while furthermore helping you to beautify your mental health.

Chapter 6: Mindfulness

The present day international is in fact a busy one. Hurrying to carry out all you want to do is probably in reality difficult, so it is no marvel that more people find out that we've were given in truth out of area our reference to the prevailing 2nd. A amount people find out that we're dropping out on the satisfaction of the minute. We forget about how we are feeling at any time and this could reason detrimental repercussions physical and psychologically in our lives.

Did you upward push up feeling rested in recent times? Did you have got a have a examine those plant life flowering to your street these days? Did you pay hobby the birds making a song as to procure to paintings? If the response to the ones have come to be no, you need to bear in mind training mindfulness.

What Is Mindfulness?

The term "mindfulness" is applied to provide an reason for the workout of focusing all your hobby deliberately at the minute you stay in and accepting your emotions and sensations and now not the use of a judgement. It has truely been shown to be a enormous element in accomplishing famous pride and lowering anxiety stages.

Mindfulness has its origins in Buddhism, Nevertheless, nearly each faith consists of a few shape of prayer or meditation method to transport your perception patterns far from standard fixations and closer to gratitude of the prevailing 2d.

Practicing mindfulness has genuinely been established to carry a wealth of upgrades to each physical and intellectual signs, assisting in bringing favorable change to mindsets, behavior and health. When you're aware, you can take pleasure in existence's moments once they seem. This permits you to interact extra absolutely with sports activities sports and help you in coping better with unfavorable sports to your lifestyles.

If you consider the now, you may have a minimized possibility of having caught up at the side of your problems approximately topics you've got were given have been given genuinely completed to your beyond or subjects you could carry out within the future. You'll have much less fixations with fulfillment and self-self notion while moreover having the capacity to form deeper and higher connections with exclusive people.

Mindfulness has definitely been established to decorate your physical fitness. It can assist in enhancing coronary coronary coronary heart health, disposing of anxiety, decreasing your immoderate blood strain, lowering discomfort, improving your sleep and even reducing intestinal problems. On the possibility hand, it gives a bunch of mental health advantages collectively with the consolation of substance abuse, depression, anxiety, consuming disorders and OCD.

How Does It Work?

Specialists assume that, in part, mindfulness works by assisting in permitting humans to clearly be given emotions and reports in location of responding with hostility or avoidance.

Mindfulness might be practiced in hundreds of strategies., Nevertheless, the primary objective of mindfulness techniques is to grow to be being more alert and centered and unwinded via paying very close to hobby deliberately to the thoughts and emotions you enjoy at any given minute without judgement.

Consequently, your mind can refocus correctly on nowadays.

There are numerous mindfulness techniques. Nevertheless, that could be a huge guide to which incorporates mindfulness practices into your lifestyles:

- Start with the useful resource of taking a seat silently and concentrating to your respiration styles. Additionally, you may attention on a phrase or mantra that you restate to yourself quietly. Allow your thoughts to transport without judgement, returning to concentrating on your mantra or respiration.

- Note the diffused research to your frame like itching or tingling. Once another time, do now not compare them, absolutely allow them to skip. Concentrate your consciousness on each part of the body out of your head on your feet.

- Note the noises, tastes, factors of hobby, smells and touches round you. Once all another time, with out judgement, honestly permit them to come again and circulate.

- Handle your cravings, whether or not or not they'll be for a sample of conduct or for a compound. Acknowledge the sensations, but, permit them to go through you with out judgement.

You can begin training mindfulness by myself via the usage of tai chi, yoga or different meditation techniques. You genuinely need to enlarge attention, check the sensations and thoughts streaming via your frame without judgement, and notice the reports which you experience. With time and exercise, you can discover that you grow to be being greater thrilled and similarly self-aware.

Tips for Introducing Mindfulness Practice into Your Life

If you are organized to offer mindfulness exercise into your lifestyles, you may possibly start taking part in a category or buy a meditation CD to begin operating towards. There are, despite the fact that, plenty much less traditional techniques you could include. Here are a few number one thoughts to help in breaking the ice to mindfulness:

- Select an interest within the route of which you may engage in mindfulness. It is probably whilst you're taking walks, ingesting or showering.

- Begin thru concentrating on the feelings your body is experiencing.

- Breathe gradually thru the nose. Permit the air to move downwards in your lower stomach and permit your belly to completely growth.

- Breathe frequently out via the mouth.

- Consider the feelings you revel in whenever you breathe out or breathe in.

- Gradually, keep together with your interest with attention.

- Totally have interaction all your senses. Consider every touch, sight and noise. Relish each experience.

- If you recognize your thoughts is straying from the prevailing hobby, deliver your attention again cautiously onto the sensations you're experiencing.

Chapter 7: Whole Foods

Although anyone recognize that we need to consume extra entire ingredients, it can be all too appealing to attract on the processed meals that we find out in eating places and shops all over. Fast food is an ever-gift characteristic of lifestyles inside the worldwide in recent times.

The life of McDonald's, KFC and Burger King in buying centers anywhere inside the america of the usa is simply motivating extra people to snack on that from an early age. More people nowadays are consuming processed meals than ever within the beyond, but, the exchange-off for ease is a bunch of health problems for each the mind and body.

Why is dangerous meals such an trouble, and the manner should it have an poor affect for your health? Here, we take a more in-depth take a look at why whole components are a much better choice for your regular healthy eating plan and the manner you can gift them higher into your existence.

The Unhealthy Food Issue

The definition of dangerous food is meals that is inadequate in nutrients and thick in calories. Over the past couple of years, comfort and junk meals consumption have sincerely appreciably prolonged, and nowadays, approximately a quarter of the population ordinarily takes in processed meals. Consequently, there has in reality been an growing epidemic of persistent infection.

The primary trouble associated with the consumption of unhealthy food robotically are weight issues. It is expected that the load hassle rate in 2050 inside the U.S.A. Alone with reach forty two percent. Kids taking in processed food often consume greater carbs, fats, and processed sugars and lots a great deal much less fiber than they require. Consuming 187 more electricity day by day than they want, it is now not surprising that 6 pounds of weight is located on each year, developing their odds of organising coronary coronary heart disease and diabetes, to call a few chronic ailments.

Another hassle growing from processed food utilization is the chance of organising diabetes. Insulin tiers growth whenever you take in processed sugars which can be decided in white flour, sodas and other food which does not have the important vitamins and fiber to metabolize carbs efficiently. If you devour unstable food at some stage in the day, your insulin ranges can emerge as being chronically excessive most important to insulin resistance gradually. This triggers kind II diabetes to reach.

If you dispose of fiber, nutrients and minerals from your eating plan, you may become being nutritionally lacking. This can cause low strength tiers, sleep disruptions, low performance, and country of thoughts swings. High stages of salt positioned in processed meals furthermore result in the overconsumption of salt. This consequences in coronary heart, liver and kidney contamination along facet excessive blood strain.

Not all the troubles arising from using volatile meals are physical. Some are mental too. A 2015 research observe mounted that humans on excessive glycemic eating plans suffered greater from depression than folks who had a low GI ingesting plan. Because junk meals-heavy eating plans are so bad for us, it stands to reason that we want to try to find a far higher eating plan that promotes health and well- being. This is in which entire meals can enter into play.

What Are Whole Foods?

This time period is utilized to give an explanation for meals which is nearest to its natural u . S . A .. They gain us thinking about they create more nutrients than packaged and processed food. Professionals suggest that we want to all be going for entire food comprising spherical 75 percentage of our normal ingesting plan. This is going to help us to live healthful, without infection, with slower getting older.

What substances ought to we be consuming?

Whole substances embody stop cease result and veggies which have now not been processed on the side of complete grains like millet, oats, quinoa, buckwheat, cornmeal, rye, and wild rice. We need to additionally be consuming extra beans and beans like lentils and chickpeas together with more nuts and seeds. Wholefoods moreover encompass the ones stemmed from animal origins which incorporates eggs, fish, rooster, seafood and lean pork like veal, beef, red meat and lamb.

If you eat unprocessed elements, you will have the capability to take in the remarkable amount of day-to-day vitamins you want for widespread health and fitness, and in the very satisfactory viable opportunities.

Wholefoods embody severa vitamins all in a unmarried meals which encompass vitamins, minerals, fiber, vital fat, and phytonutrients. They are moreover in truth enough in materials which can't be synthetic in the body and which, because of this, ought to be obtained with the aid of your ingesting plan. For instance, valine, an amino acid, can't be made thru the frame itself, and for this reason, desires to be supplied by manner of the usage of what you consume. It is critical for tissue restore art work and muscle metabolic methods, so incorporating loads of complete meals to your normal recurring is important.

When you devour entire factors of their herbal state, you could take benefit of the synergy end end result of the vitamins within the food interacting to gain your body's healthy performance. For instance, tryptophan, an amino acid, goals B vitamins to turn out to be serotonin. Additionally, complete food are huge inside the anti-oxidants which reduce the effects of loose radicals and conflict troubles like most cancers and cardiovascular sickness.

One More Reason to Eat Whole Foods

For severa years, specialists have sincerely been informing us that quit end result and greens are essential for our well-being. Nevertheless, a whole lot humans however discover it tough to encompass sufficient of them into our ingesting plans. Yet, entire food can save you us from wearing out being ill and help in preventing the problem of weight issues.

Lots of research research have certainly uncovered that consuming more complete food goes to offer your body with essential vitamins sources collectively with calcium, fiber, magnesium, B vitamins, protein, Vitamin D, potassium and critical fats which guarantee your frame's cells artwork in the proper way. Foods which can be processed are tough to take in effectively and might make you revel in tired and unwell.

When you embody more entire meals into your existence, you could revel in a group of benefits along with:

- Enhanced blood glucose levels. Processed meals consist of insulin growth issue that makes your blood glucose levels more. As a quit result, you enjoy blood glucose swings and yearnings. Entire elements will not create the ones spikes and are going to let you hold balance throughout the day.

- Enhanced meals digestion. Entire meals encompass extraordinary offers of fiber, that may be a vital nutrient for meals digestion. This fiber is natural and goes to will permit you to experience fuller for longer on the same time as moreover assisting your food digestion and lowering your blood glucose levels as it breaks down regularly inside the body.

- Greater electricity levels-- the body is more capable of obtaining energy from healthful elements than processed ones, so you'll start to sense more invigorated with a faster metabolic manner at the same time as you eat more whole components.

- Minimized soreness-- processed foods have immoderate inflammatory houses. Given that they may be acidic in nature, they expand pH degree imbalances that can bring about continual ache scenario symptoms and signs worsening. Entire elements keep your body greater alkaline, and due to this, with out swelling and ache.

How to Consume More Whole Foods

Do you want the advantages of consuming extra entire components, however, do now not understand the manner to encompass them for your consuming plan? Here are a couple of rapid mind to aspect you in the appropriate course:

- Change to conventional oats instead of immediate oat cereals. Instant oats are going to normally have oat bran disposed of. This technique that a first-rate deal of the vitamins and fiber have surely been disposed of, decreasing its nutritional in reality really worth.

- Change to finish fruit and vegetables in preference to packaged juice. When fruit is juiced, it subsequently ends up being a centered sugar source, and this increases your blood glucose degree an extended way extra rapid at the same time as in comparison to whole give up result. Juicers moreover get rid of the pulp and pores and pores and skin of the fruit, so flavonoids and anti-oxidants are removed. Packaged juices moreover have more sugar mixed with chemical compounds and preservatives.

- Change to clean fish as opposed to canned or frozen fish. Fish encompass essential fats that are generally disposed of or minimized in some unspecified time inside the future of the product packaging method. You require omega 3 fat to hold your involved, immune, cardiovascular and reproductive structures working efficaciously.

You can speedy find out complete materials to be had on the market in grocery keep in case you buy groceries inside the aisles dedicated to clean meals. As you circulate closer to the middle of the shop, go through in thoughts that you can find out greater processed components, so try to buy groceries on the out of doors edges of the shop. You'll moreover discover entire ingredients in farmers' markets and at herbal meals shops.

Chapter 8: Probiotics

The majority people generally tend to accept as true with that germs are lousy and perilous for us., Nevertheless, this isn't the case. There are numerous sorts of germs which might be beneficial for our our our bodies. These live micro-organisms are referred to as probiotics, or nice germs, and might help in making your frame masses extra healthy.

There are numerous kinds of probiotics that supply numerous benefits in your properly-being and fitness. They perform in the GI gadget to enhance the immune tool, stopping risky germs from ending up being linked to the indoors wall of the intestinal tracts whilst improving the stableness and feature of the intestinal lining's natural microflora.

Typically, the human frame has an notable stability of micro organism, but, there are precise medical factors that might bring about imbalances. Consequently, the sickness-inflicting germ numbers can develop extensively. Unneeded use of antibiotics, intestinal problems, surgical remedy, taking PPIs, persistent strain, sensitivity to gluten, or even the everyday American diet plan can motive such imbalances.

Fortunately, there are techniques to treatment the stability of bacteria for an awful lot higher present day day fitness and health. The very fantastic approach is to function probiotics in your each day eating plan.

What Are Probiotics?

Probiotic is a term applied to demonstrate the live germs which exist in yogurt and fermented components. They can advantage your digestion tool by way of manner of turning inside the stability of the great and horrific germs in your microbiome in alignment. This ensures you have a decrease danger of experiencing diverse clinical ailments and conditions.

Probiotics may be crafted from health food resources like yogurt and kefir, but, they could likewise be crafted from meals that have virtually been superior with probiotics alongside element from expert nutritional supplements. It's typically great to get your probiotics from healthful meals belongings, notwithstanding the truth that.

Probiotics take severa shapes and can be decided in these components:

- Sauerkraut and unpasteurized kimchi.

- Miso soup

- Soft cheese and enriched milk.

- Sour pickles in saltwater

- Sourdough bread

Here are most of the most beneficial probiotic meals that you can take pride in each day:

- Yogurt-- that is a major probiotic source as it includes milk which has absolutely been fermented through best germs like bifidobacterial and lactic acid germs. Not surely can yogurt growth your bone health, it could lower high blood pressure and eliminate undesirable signs and symptoms related to IBS (irritable bowel syndrome). Not every type of yogurt includes probiotics, so you must simply pick out out yogurts with active or live cultures.

- Kefir-- this probiotic fermented milk beverage is constituted of cow or goat milk with blanketed kefir grains. Once all over again, kefir can enhance the bones, resource with gastrointestinal problems, and guard the body from infections.

- Sauerkraut-- this is crafted from shredded cabbage which has genuinely been fermented with lactic acid germs. Not honestly is sauerkraut loaded with vitamins and fiber; it additionally consists of manganese, salt and iron similarly to anti-oxidants which beautify eye fitness. You ought to make certain that the sauerkraut you have got were given clearly picked is unpasteurized to enjoy its probiotic benefits.

- Tempeh-- this item is comprised of fermented soybeans and is terrific as a opportunity for meat. The fermentation manner implies that you may absorb extra minerals from tempeh. It is furthermore an full-size deliver of nutrients B12 while moreover having probiotic advantages.

- Kimchi-- this spicy, fermented Korean meal is usually made from cabbage which has virtually been seasoned with chili pepper, garlic, ginger, scallions and salt. It furthermore includes Lactobacillus kimchi which advantages your gastrointestinal fitness.

- Miso-- this Japanese flavoring is made out of fermented soybeans. Typically made into soup, miso is an super supply of protein and fiber and is loaded with plant substances, vitamins and minerals

- Kombucha-- this fermented black or inexperienced tea drink is from Asia.

- Pickles-- gherkins are fermented in salt and are an amazing supply of probiotic germs which beautify gastrointestinal fitness.

- Buttermilk-- fashionable buttermilk is the liquid that is left after developing butter. It includes probiotics similarly to crucial nutrients and minerals.

- Natto-- this fermented soybean item resembles tempeh and miso which incorporates Bacillus subtilis. It is additionally excessive in protein and healthy eating plan K2.

- Some cheeses-- even though hundreds of cheeses are fermented, they do not all encompass probiotics. Just human beings who've lively and stay cultures do. Cottage cheese, cheddar, gouda and mozzarella are all quality examples of cheeses wherein the pleasant germs make it thru the device of developing vintage.

There are severa probiotic food which you may encompass on your eating plan, though, if you do no longer like they all, you would likely constantly strive a probiotic supplement which can be taken each day to decorate your great fitness and health.

Chapter 9: Cryotherapy

Although cryotherapy has actually been spherical for a long time, it has great just presently reached terrific hobby. This is due to the reality that celebs have definitely started to proclaim its virtues, specifically sports stars who united states that it assists in enhancing their restoration time after workout.

Cryotherapy has virtually been developing as a fad in spas and health centers because of its appeal with celebs and expert athletes. Those who have this treatment take delight in quicker healing times after workout and assume that the sub-zero temperature tiers can decrease allergic reactions, anxiety and stress and arthritis, further to developing the pores and pores and pores and skin appearance more youthful.

Cryotherapy intervals can take some of shapes. The maximum commonplace encompass fame in an entire-body cryo-chamber. This chamber is a can-like enclosure with an open-pinnacle, so the purchaser's head constantly remains out. On the alternative hand, the the rest of the frame is subjected to sub-zero temperature levels.

When having WBC, you operate gloves, underclothing and socks to avoid your extremities putting in frostbite. Each session genuinely lasts some of minutes and even though it is able to probably experience a hint uneasy and weird, it is not undesirable or uncomfortable. There are moreover focused cryotherapy treatments that consist of just exposing one a part of the frame to sub-0 temperature levels. This works for relieving pain in a specific region.

The idea at the back of cryotherapy is to lower the body temperature degree to this kind of diploma that the "fight or flight" mode is released. This triggers the body to supply the blood from the extremities to the heart wherein it's miles rapid oxygenated and pumped full of vitamins. On departing the chamber, the newly oxygenated blood is pumped outwards lower lower back throughout the body all all another time to decorate restoration and restoration.

What is Good About Cryotherapy?

Cryotherapy is notion to offer some of intellectual and physical benefits. These embody:

- Decrease of migraine symptoms-- cryotherapy is idea to cope with migraines because it numbs and cools the nerves throughout the neck, cooling off the blood which goes via the intracranial vessels.

- Minimized nerve contamination-- cryotherapy is used by professional athletes to numb discomfort from inflamed nerves. It works in handling neuromas, excessive injuries, pinched nerves or chronic ache.

- Manages temper troubles-- even as the entire frame undergoes sub-zero temperature ranges there can be a physiological hormone reaction within the body that includes the release of noradrenaline, adrenaline and endorphins. This can assist folks that experience anxiety and depression to enjoy a miles higher kingdom of thoughts.

- Lowered arthritic ache-- human beings suffering with arthritis can revel in heaps lots less pain at the equal time as having both complete frame or localized cryotherapy.

- It gives with pores and pores and skin problems-- if you experience pores and pores and pores and skin problems like atopic dermatitis, you will probable discover that cryotherapy can assist in assuaging itching and dryness. Cryotherapy lowers swelling at the equal time as moreover enhancing the stages of anti-oxidants inside the blood which assists in improving the situation of the skin. For the identical reason, it's miles furthermore useful in assuaging times of zits.

- Enhanced weight reduction-- weight advantage is a large problem worldwide nowadays; however, cryotherapy has definitely been placed to have a few weight-loss advantages thinking about that it speeds up the metabolic way for severa hours following remedy. This indicates that the ones who have WBC can effectively burn greater energy after having a treatment consultation.

Chapter 10: Cleanse Your Air

Are you involved about impurities in the air? Do you conflict with allergies or allergic reactions? Do you have got have been given animals in your property, or do you live with a cigarette smoker? If the reaction to any of those is sure, you should recollect taking a examine techniques to cleanse your air. Here, we take a better observe some of the motives that more individuals are selecting to shop for an air cleanser as a primary hack for a miles greater wholesome way of life.

Why Should I Cleanse my Air?

We frequently suppose that the air inner our houses is wholesome and easy to breathe. Nevertheless, it'd come as a marvel to find out that it could be simply as infected and contaminated due to the reality the air out of doors.

Smells, dirt, animal dander, mildew and smoke are definitely the numerous pollution that you could discover inside the air inner your property and there are some pollution interior your property which exist in portions five instances better than within the air outside. It's no wonder, then, that many human beings revel in allergic reactions and allergic reactions. Discovering a manner to capture unwanted particles from the air like dirt and pollen is critical.

So, the way to determine if the air in your private home calls for cleansing? Here are simply most of the reasons to think about:

- You have animals. If you have got got had been given a cat or dog , you'll probable wind up experiencing respiration troubles. They shed dander onto floor areas of your home which cannot be gotten rid of via vacuuming on my own. Cleansing the air is the very exquisite method to be super of putting off this problem effectively.

- You experience hay fever, hypersensitive reactions or hypersensitive reactions. Throughout the summertime and spring, hay fever is a normal difficulty because of the pollen debris which is probably within the air. These get worse the eyes and might cause bronchial asthma. Cleansing the air is going to remove those irritants so you can live comfy all season prolonged.

- You have a mould problem in your home. If your house is damp or prone to damp, you may probably find out that mildew is an issue. Restrooms and kitchen regions are tough zones and with out disposing of the spores from the air, you could set up respiratory issues. Cleansing the air receives rid of this possibility.

- You have dust mites. Dust mites live in honestly all and sundry's residence and might cause allergic reactions at the pores and pores and pores and skin further to respiration issues. If you cleanse the air, you may now not want to strain over this hassle.

- You are living with a cigarette smoker. Smoke from cigarettes can cling in the air for a totally long term triggering respiratory troubles in willing people, no longer to say unwanted smells. If you cleanse the air, this may no longer be a trouble.

- You do no longer like cooking smells. Whether you live close to next-door buddies who prepare robust-smelling components or whether or not you prepare them yourself, you do no longer need the smells final around your home. An air cleanser can do away with the unwanted smells.

- You have jeopardized immunity. Airborne infection debris can go along with the float in amongst humans after they sneeze or cough. If you cleanse the air internal your private home, your complete family is going to have masses better fitness and stylish well-being.

- You have an toddler. Young kids are specifically vulnerable to air-borne bacteria and viruses. They can also moreover be greater at danger while they will be exposed to risky pollutants and contaminants in the indoor environment. Cleansing the air goes to provide your kids the very extremely good opportunity of taking pride in best fitness.

- You stay near to a roadway or a farm. If you are dwelling in a vicinity this is at excessive threat of infection, you have to cleanse the air internal your private home to maintain the hazard of infection to a minimal.

As you could see, there are numerous reasons for thinking about cleaning the air in your property to beautify your common health and fitness.

How Can I Keep the Air in my House Pure?

There are various of factors you can do to maintain the air in your home herbal and tidy. Among the very remarkable is to buy an air cleaner. This is a device that has numerous filters similarly to a fan to absorb the air and distribute it whilst catching the contaminants and particles and pressing tidy air decrease lower returned to the home.

When choosing an air cleanser, you want to make certain to choose one that has a HEPA smooth out (excessive-performance particle air filter out). These seize debris of a sequence of sizes in a sincerely fine multi-layered web made from fiberglass threads. This airtight filter guarantees that even the smallest extraordinarily-fine particles are caught in order that they can't be launched into the surroundings to cause problems. You ought to pick out out an air cleaner this is massive sufficient to successfully smooth up the air in the vicinity that you are dwelling in, and which has a tidy air shipment charge of over 350. This goes to make sure that the air stays as pure and easy as possible.

There are moreover a few precise topics you may do to beautify the incredible of the air internal your own home. For a begin, even though it might possibly sound counter-intuitive, you may keep the home windows open, producing a pass-draft on every occasion feasible via the use of setting up the window on opposite sides of your regions. This goes to assure that unwanted contaminants and pollutants will no longer come to be being stuck interior your private home, triggering your health issue.

You need to moreover vacuum your floors frequently to put off dust mites that could cause pores and skin and respiration issues, and make use of exhaust fanatics in your laundry places, restroom and kitchen place to prevent mildew from developing and triggering respiration problems and important illnesses.

You have to stay clean of lights a wooden fireplace inner your own home and prevent cigarette smoking inner your house as that is going to help in improving the extraordinary of air interior your house too. Smoke can cause respiration issues whilst secondary smoke from tobacco may be clearly unfavorable for your health, even triggering most cancers every so often.

Obviously, you have to additionally maintain in mind to regularly adjust the filters on your air cleanser, vacuum and air waft device to ensure that the air inner your private home stays smooth and healthful. A blocked filter out can't consequences lure particles and pollution, so that you want to make sure to stay on top of the converting schedule to make sure that most contaminants are gotten rid of from the air that you and your own family breathe every day.

Chapter 11: All you want to recognize approximately biohacking

The very first time my little sister positioned "Biohacking" at the internet, she had a superb antagonistic approach to the idea. She went on approximately how hackers in any project are criminals certainly searching out shortcuts to the whole thing. However, what maximum human beings fail to apprehend is that, there are or more elements to the whole lot, For instance, enhancements in technology and engineering has introduced approximately the improvement and stream of modern-day gadget and tablets that have stored quite a few lives, on the other hand, it has furthermore contributed to the invention and usage of nuclear bombs and guns of mass destruction that has killed plenty and plenty of human beings.

Biohacking consists of sporting out organic engineering hacks commonly to enhance the lives of people. Although some human beings go to immoderate lengths, carry out uncensored experiments and make use of all way of unorthodox concoctions, distinctive humans ought to even bypass as an prolonged manner as the usage of biohacking for the motive of hurting others. But this does not alternate the reality that Biohacking ought to provide treatment to numerous ailments in the nearest destiny, and certainly in all likelihood all you want to do the reputedly now not viable and acquire your set-desires is to bio-hack your subconscious thoughts. So, get snug, open up your thoughts, as you are about to find out the fastest manner to bio-hack your subconscious thoughts.

What is Biohacking

Biohacking is largely defined due to the fact the attempt to enhance your thoughts and manipulate your frame which will optimize popular standard performance, outdoor the arena of traditional remedy. There are particular various definitions for biohacking because of its significant scope. The idea modified into postulated thru a small group comprising specially rich those who desired to find approaches to optimize and improve their minds and our bodies. Ever considering humans have engaged in DIY bioengineering and nevertheless do till in recent times.

Today Biohacking has grow to be a international phenomenon, evolving from smooth bio-hacks like fasting, meditation and hypnosis to putting the opposite way up(supposedly imagined to hack the thoughts with the resource of growing blood drift to the mind), Nowadays, bio-hackers take delight in more drastic practices, examples of which may be indexed under:

• Near infra-pink saunas (Expected to relieve pressure through electro-magnetic transmissions)

• Neurofeedback (Training to display display screen your mind waves)

• Cryotherapy (making your self cold deliberately)

• Virtual go along with the go with the flow tanks (meant to spark off a meditative united states via sensory deprivation)

• Young blood transfusion (speculated to gradual the developing older method or maybe contrary it), fecal transplants among others

• Fecal transplants (entails the switch of feces from a wholesome donor to an volatile recipient)

There are many extra Biohacking practices; maximum are aimed toward optimizing body functions. The Biohacking community considers "Control" as their watchword. Additionally, it is not uncommon to find out biohackers utilizing various capsules like anti-depressants and various nutritional dietary dietary supplements like anti-developing vintage nutritional dietary supplements, nootropics, and a whole lot greater. A subdivision of biohackers called grinders go to excessive lengths as implanting microchips in factors of their bodies, for diverse functions as monitoring blood sugar stage subcutaneously, or tracking. The "grinders" accept as true with that those pc chip implants are beneficial and could result in a better lifestyles for humans living with sure disabilities. For example, someone who has misplaced the use of their hands have to have a laptop chip implanted of their ft, and ultimately have a chip scanner established of their doors in addition to superb family objects across the residence, for ease of operation.

However, Basic math teaches that a proper away line is the shortest distance among factors, so certain, that inner voice telling you that short reduce on occasion lessen you brief, couldn't be extra proper. The truth is short cuts maximum times won't be as short as they seem, if you are going to interact in biohacking, you need to be properly acquainted with what you are going into. It is usually recommended to workout tested bio-hacks due to the reality some bio-hacks are quite risky, I mean it makes no revel in to die at the same time as in search of to elongate your lifestyles, besides if your scenario end up terminal and you've got have been given now not whatever to lose.

Bio-hacks inclusive of CRISPR, fecal transplants, greater youthful blood transplants aren't authorized via FDA, because of the reality they'll be not considered safe, as some people react negatively to those hacks some even come to be six ft beneath. However, most mind hacks are constant to workout; in fact, studies have demonstrated that meditation can assist lessen ache and tension.

You may need to invite yourself, is biohacking truly nicely well worth it? Because of the dangers related to converting human biology, what is more, biohackers intend to engage in more risky hacks to understand extra widespread effects. They agree with we're capable of engineer our way beyond our our bodies shortcomings, and over the years evolve into an upgraded species, more potent and smarter with higher abilties, a global of incredible-people and immortals. If all we want to do is to make some experiments and sacrifices to get there, then it's a honest deal.

The Effects of Biohacking on Your Subconscious Mind

Some days we want we should simply interest amidst chaos, in that on the spot attention appears like its slight years away. At the very 2nd you need to get your act together and get some artwork achieved, fatigue sets in and you discover your self seeking to study the equal paragraph again and again again. Getting our mind and body to feature at their notable all of the time is not an clean project, our very personal body appears to fail us at very important moments, however its everyday despite the fact that, ultimately we are most effective human, sure to err. Yes, it's normal to interrupt down, and yes, we are first-class human, but our obligations and commitments require super-human capabilities. You have to lodge to caffeinated beverages, or a few weight loss plan B pix, but the ones aren't so powerful each. They assist you live alert for some time, but you become all jittery and stressful and it doesn't even assist you concentrate. Instead of giving up and bickering about how "extremely good-annoying those bosses are", it's miles excessive time you be part of the bio-hack train.

Most human beings are typically skeptical about "turning into a member of the bio-hack train", and I constantly tell them, "we commonly have a tendency to worry what we don't recognise". Before you finish which you don't need to bio-hack your thoughts, you want to have great statistics of what it is approximately and what it could do for you. Biohacking essentially consists of the use of some detail available to you (this may be remedy, remedy or maybe computer chip implants) to optimize the general overall performance of your thoughts and body, Biohacking doesn't always advise you need to visit excessive lengths; it is able to be as smooth as meditation or adjusting your eating sample.

Biohacking the subconscious mind addresses loads of issues, a number of which is probably noted below. Some effects of biohacking on the subconscious mind encompass:

1.Increased I.Q: You can turn out to be a genius in case you want to. We have been made to really receive as actual with the stereotype that one's diploma of intelligence is constant and cannot be improved. This idea is certainly a false impression; in fact, "David Asprey" the daddy of all element's biohacking has upgraded his mind via extra than 20 I.Q points. So, no longer anything stops you from turning into the following Einstein in a new global in case you so please.

2.Increased Attention: One vital place Biohacking your subconscious thoughts offers you is prolonged interest. It allows you to recognition for an extended time period, without getting distracted. This allows you to pay attention on essential subjects, consequently growing productiveness.

3.Better Memory: Biohacking permits you discover techniques to speak thoughts, thoughts and memories among your conscious and subconscious mind. You growth the capacity to tap into and get better records out of your subconscious mind.

Chapter 12: How to Become Unstoppable When Biohacking Your Subconscious Mind

Apart from the hacks and strategies, the Major problem you need to be unstoppable even as Biohacking your subconscious mind is the "Super-human spirit". Once you recognize that you truely very personal the electricity to be fantastic, there may be no stopping you. These five points will help you:

1.Have The Can-do Mentality: Your chances of actually accomplishing something you receive as true with you are not capable of executing are very narrow. Therefore, the primary element to do at the same time as in the approach of Biohacking your unconscious is to have a mentality which you are able to doing the more-regular. Most university college students that graduate with variations, a good buy greater than art work more tough keep in mind that they're going to graduate with a distinction. The moment you're taking delivery of as real with you could actualize your dreams; your subconscious mind starts offevolved identifying methods to make that display up.

2.Surround Yourself with Positive Reinforcement: Motivate your self through way of surrounding yourself with extremely good reinforcements, a few aspect should energize you and spur you in advance closer to attaining your purpose should be all you allow to get to you. Keep buddies with like-minds that permits you to cheer you on, now not folks who will tear you down. Get motivational messages and songs, play them all the time. You may be surprised via how inspired you'll be.

3.Speak Positive Affirmations: Words are effective, a long way more powerful than you may remember. Mainly due to the fact the terms you communicate without trouble diffuses into your unconscious, and your unconscious responds through developing only styles that align with the ones phrases. What's greater the moment the unconscious registers the terms you communicate, it most effective sees things from that attitude, and consequently, it'll most effective pick out with activities and thoughts steady with that thoughts-set. This is why it's miles very critical to speak superb phrases best. When you try this your unconscious mind registers a awesome thoughts-set and identifies with effective mind, thoughts feelings and memories most effective. With this in thoughts be very aware about the stuff you say, say handiest what you need to mention, due to the fact what you're announcing is what you get.

four.Identify and Discard Contrary Thoughts: The significance of staying extremely good cannot be over-emphasised, make sure which you perceive and discard opposite thoughts. You can write them on a bit of paper and burn them if it makes you allow glide without troubles. Contrary thoughts make it tough to maintain without delay to the "can-do mentality", they make you experience miserable and incompetent, now that isn't accurate. One suitable manner of discarding contrary mind is to counter them with top notch thoughts. For instance, if you feel unsightly and fats, counter it with a exceptional concept, say to yourself " I am so lovely, damn examine those cheekbones and whole lips, every body could be lucky to have me." eventually, make sure you're intentional and passionate even as you say it, in advance than you recognise it you may enjoy and look lovely.

five.Make Your Future Plans a Present Time Reality: The trouble some of human beings have is procrastination. It is an evil that makes you unproductive and clearly vain at the end of the day. The unconscious mind is more privy to the winning, so in case you without a doubt intend to gather a fine motive, then you definitely definately are going to have to make it a gift-time truth. For instance, in case your purpose is to paintings on a specific project. Instead of saying "I am going to art work at the challenge" say as an alternative "I am strolling on the mission". The former statement sounds futuristic so the subconscious thoughts won't do whatever about it. The latter declaration, however, feels like a now aspect, like an ongoing occasion, that way your subconscious mind sees it as a present-time event that desires to be finished after which receives to paintings on innovative strategies that will help you acquire your set-aim.

One very important factor to be conscious is that Biohacking does no longer come up with extra brainpower, instead, it permits you to maximise the simplest you have already got, more like upgrading your android walking device. Thus, to come to be a pro at Biohacking your subconscious mind, you need to put together to alternate a whole lot about you. This includes developing a "me-time" for meditation, converting your weight-reduction plan and eating patterns and different mind bio-hacks to be stated in next chapters of the ebook.

Keep Reading!

Chapter 13: The unconscious thoughts take heed to your subconscious

The unconscious is that part of the mind which controls all additives of your lifestyles without your recognition and over which you do now not have manipulate, it operates a powerful device that runs all your life's affairs. The famous saying that human beings do not use all of their brains, but first-class a part of it's far more or heaps less due to the subconscious thoughts. You can take into account the subconscious thoughts as a databank, it shops statistics of the whole lot which encompass the ones your aware thoughts cannot accommodate. It holds your beyond research, the entirety you've got got ever seen or finished, and each sincerely one in all your recollections collectively with prolonged-time period reminiscences (Memories lengthy forgotten are buried within the subconscious).

The unconscious thoughts is answerable for loads of things, one of this is the processing of information. Your mind gets hundreds of MB of records in line with second, it would definitely explode if it had to analyze and machine all that facts. So, the subconscious thoughts steps in, video display units and strategies all of the information, and relay exceptional the crucial records to the mind. Just like a wild monkey, the subconscious thoughts have to paintings all day without tiring. You can take benefit of this unusual capability, permitting your subconscious thoughts to take a look at records for dangers and possibilities and relay such statistics to the aware mind, on the same time because it additionally receives busy with carrying out your set-dreams. Learning a manner to successfully transmit statistics from the unconscious to the aware thoughts and vice-versa lets in for a unfastened flow of records vital on your success in life.

Communication between the conscious and the unconscious thoughts is a -way technique, i.E. Your subconscious mind can communicate along side your aware thoughts and vice versa. Thus, whenever you do not forget a reminiscence or event from the past, this is your subconscious mind speaking at the side of your conscious mind. The conscious mind, but, communicates with the unconscious mind thru the precept of autosuggestion. Autosuggestion is essentially the system by way of the use of the use of which facts from your aware thoughts is delivered into the unconscious mind. This isn't as clean as unconscious to conscious communique due to the truth the thoughts want to be conveyed collectively with sincere emotion, both incredible or horrific i.E. Best mind sponsored up by means of real emotion make it to the unconscious mind.

Learning the artwork of speakme thoughts out of your conscious mind to the subconscious is a extraordinary gain due to the truth this shows you get to determine the thoughts that dominate your subconscious thoughts, and eventually, supply purpose to the internal voice. This generates excellent vibes and strength. But this may only be completed with the beneficial aid of allowing more great mind waft in while getting rid of horrific thoughts. And how do you do that? First matters first, you start through eliminating horrible mind alongside facet the negative feelings, please be conscious that that is very IMPORTANT as terrible thoughts and feelings are commonly stronger than super ones, so as that they've a bent to dominate. Be positive to again this up right now with excellent speak and be intentional and emotional about it. Remember that your subconscious thoughts policies your existence, if the internal you are loaded with bad thoughts, you emerge as getting an entire load of lousy studies on your day. If you already fear that you are going to have a shitty day at artwork, or be unproductive or fear no person likes you, frequently these types of items begin coming to bypass, because of the awful electricity you're going to begin to emit. Your journey to a wealthy and quality existence starts offevolved from your functionality to talk powerful thoughts and feelings on your unconscious thoughts, that way your unconscious is free of awful strength and without problems identifies brilliant possibilities and methods with the aid of way of which you can maximize them.

How the Subconscious Mind Operates

Over the years, human beings have made series of tries to knowledge the mind each the conscious thoughts and the subconscious, no matter the fact that there can be still lots left undiscovered, so far we recognize that the subconscious mind is that part of the thoughts that controls every of our lifestyles sports activities independent of the aware thoughts, it also includes at artwork even if we sleep. There were instances in which human beings wake up from an extended coma narrating the subjects, they witnessed and professional on the equal time as in coma. This portrays how effective the unconscious mind is and is even capable of an lousy lot extra than we're capable of don't forget.

For you to benefit your set-desires, turn out to be the wonderful model of you and gain the goals you have set via your conscious thoughts, you want to align the ones outside conscious goals with your inner unconscious mind. The only and excellent way to align your conscious and unconscious minds is to discover ways to art work with the internal you in place of art work in competition to it.

However, operating consistent with the most effective pressure in the human nature; the unconscious mind calls for an intensive expertise of procedures it operates. Your subconscious thoughts operates in a extraordinary manner from your aware thoughts. It follows high satisfactory guidelines that range appreciably from those your aware mind can relate with. Hence, operating constant together with your conscious thoughts desires which you recognize what the ones recommendations are, apprehend them and understand the manner to workout them on your every day dealings. Since the ones policies comply with to everybody, analyzing about them enables you to relate properly with other humans, no matter the form of dealings they've with you.

In this e-book, twenty-4 essential regulations of the subconscious mind were recognized, based totally absolutely mostly on works from splendid masters of the subconscious thoughts whose works are based totally on years of thorough research at the techniques, functions, language and nuances of the subconscious thoughts. The first 8 tips had been based totally mostly on the works of a master hypnotist; the late Charles Tebbets at the same time as the subsequent 16 based totally on the works of Dr. Sherry Buffington the originator of the RAMP and AMP strategies. These are the 24 Rules:

1.Every idea or concept reasons a physiological reaction: As stated in advance, every perception is typically related to emotions, which could each be extremely good or horrible. It was moreover stated that your unconscious thoughts controls certainly the whole thing approximately you along with your emotions. Consequently, which means that that your unconscious thoughts comes to a choice in case you are satisfied, unhappy, irritated, indignant or harassed. It may be instrumental to the diploma of splendid or awful energy you experience around you, is accountable for making you experience unstoppable and on top of the arena or depressing and crashing to rock backside, your feelings are that effective, they've an impact for your physical reactions greatly. Thoughts which is probably subsidized or associated with robust emotion, makes its way on your subconscious and imprints itself at the partitions of your mind, and the moments those mind are allowed to settle, they create effective thoughts, such mind glide right away to generate physiological reactions, and continues generating the ones reactions, non-stop production of these reactions ought to cause distress or perhaps issues if there is battle among those reactions and our idea of self and moreover our dreams. This manner that such conflicts may want to bring about highbrow and physiological illness; you may fall ill or be healthful, depending at the united states of america of your unconscious. In order to eliminate horrific physiological reactions, you want to discover ways to get right of entry to your unconscious mind to remove terrible emotions and mind liable for such response.

2.What is Expected Tends to Be Realized: The conscious mind without problems recognizes records transmitted with the aid of way of the use of the subconscious mind. The mind handiest responds to pix printed on the subconscious mind each those from outside assets or self-added on stimulus. Likewise, as soon as highbrow images are original, they invent a sample which the unconscious thoughts follows constantly. These styles are identified and repeated thru the aware mind, because of this reflecting for your each-day dealings. This stresses the importance of always maintaining a extremely good kingdom of thoughts.

three.Imagination is More Powerful than Knowledge When Dealing with the Mind: Ever perplexed why it's also hard to motive with a enthusiast, regardless of how logical your trouble is, they may by no means take transport of as genuine with you? The motive is that any concept or perception that is followed with the resource of robust emotion like love, hatred or anger is deeply imprinted inside the unconscious mind and more often than not can't be changed with the aid of looking to cause it out. It is vital to generally keep in mind this rule.

4.Opposing Beliefs can't be held on the same time: Having a right facts of this rule makes it much less complicated to understand how the subconscious mind operates. The subconscious thoughts seeks congruence, so at the equal time as it's far furnished with conflicting thoughts it can outstanding get preserve of one in every of them. The subconscious mind great identifies with comparable thoughts so while opposing mind are provided, it could gather only in reality clearly certainly one of them. This is why there are such severa battles are fought over variations in ideologies and belief structures. When someone starts offevolved offevolved to simply accept as proper with in a few element and is captivated with it, mainly in a few component greater than oneself, such man or woman starts offevolved to form alliance with others with comparable perception as his. A conventional example of that is the diploma of unrest and wars amongst spiritual sects.

five.Once a Belief or Idea Has been mounted with the aid of manner of the Subconscious Mind, it remains until it's miles changed thru way of Another Belief or Idea: When a perception lingers inside the unconscious mind for a long time it progressively turns into a addiction. This is the method via which behavior are shaped, each the best and horrible behavior. A consistent dependancy bureaucracy a sample over the years it truly is displayed every time the pattern gets caused. Rules 18 and 19 explains this rule considerably.

6.An emotionally precipitated symptom has an inclination to create herbal exchange if continued in long enough: The unconscious thoughts is resident within the body, and so each have to exist in unison. The frame responds to mind and mind that the unconscious dwells on. This implies that during case you stay on fear, say you are commonly frightened of ill-health, it is best a rely of time until your body tool adjusts to your idea patterns.

7.Each idea acted upon creates lots less competition to successive tips: Newton's 0.33 law of movement says gadgets in movement typically usually generally tend to live in motion besides acted upon with the useful aid of an outside pressure, this regulation moreover applies to mind travelling the neural pathways of the thoughts. When a pattern has been mounted, the body continues to carry out that addiction whenever the sample is added about. Hence even as small commitments are completed efficiently, there are better chances that subsequent ones can even come to be a fulfillment, even large commitments. This validates the adage "fulfillment breeds achievement" and "failure breeds failure".

eight.When coping with the unconscious mind and its talents, the greater the conscious attempt, the lesser the unconscious reaction: This is the motive we are saying will-strength does no longer really exist. The extra you try to combat off some matters, the more not possible it seems. For example, the extra an insomnia affected person tries to sleep, the more wakeful he becomes. The equal applies to addicts, the greater they may be trying to prevent searching the addictive substance, the more they are attempting to prevent the use of the addictive substance, the extra they want it. This is real of human beings attempting to triumph over an addiction; the more hard they may be attempting not to need the addictive substance, the more they need it. It is high-quality to deal with the problem from the inspiration reason and no longer the effect, and the reason of all we do, and experience is inside the unconscious thoughts.

9.The subconscious mind has its private language: The unconscious mind does no longer communicate with the language your aware mind does, its language is photos, feelings and sensations. It absorbs the world spherical you in the shape of a story; it takes a holistic view of factors, no longer leaving out any element. You recognize how you experience at the same time as you endure in thoughts a dream so in reality, it nearly feels as real as in spite of the truth which you had been huge awake and wasn't dreaming in any respect. The factors of interest, sounds, scent, and taste are each bit as real as your waking lifestyles.

10. The subconscious thoughts does now not differentiate among what is actual and imagined: Say as an instance, you were requested to count on your self sucking on a lemon, now, in reality there's no lemon in sight, but your body may but react to the creativeness as no matter the fact that you were truly sucking on a lemon. You have to honestly start to experience the flavor of the lemon on your mouth. You generally tend to get the same response with or without the I can preserve in thoughts.

eleven. The subconscious mind is illogical: The unconscious thoughts, no longer just like the conscious in not logical. Rather than trying to make feel of factors to look if they'll be certainly logical, it honestly connects carefully related information to attain at conclusions which can be quite everyday and complex. For instance, a depressed person, may consciously attempt to discover happiness, and within the absence of mind that could in reality make them satisfied, the unconscious mind can also need to complete with the idea that lack of existence equals happiness, and such someone ends up committing suicide. Another example is the continuing software program of mind like "you are not grown sufficient to do that all via way of yourself" even as soon as we already private the capabilities, that might have made discarding the concept an affordable detail to do.

12. The subconscious thoughts does not purpose: The instance above in Rule 11, explains this successfully.

thirteen. The subconscious thoughts does now not filter statistics: The unconscious mind best accepts records. The aware thoughts is answerable for filtering records.

14. The subconscious mind does no longer recognize non-seen terms: As defined in rule nine the unconscious mind communicates with pictures, that is why whilst you communicate with people your mind has a bent to picture the terms with a visual thing; at the identical time because the opportunity terms sincerely seem to help you make greater enjoy of the idea. For example, even as you pay attention the word movie star, your mind doesn't see the phrase movie famous person, it photographs the picture of a movie celebrity, in case your preferred celeb is Beyoncé, and you will probable see her. While other phrases that lack visual thing help your conscious mind create a linear series that will help you make more feel of the concept.

1.The unconscious mind does no longer recognize negatives: Since the unconscious mind operates on seen content cloth fabric, and terrible phrases like no and not lack visual component, that is why we fail to accumulate what we need every so often. So, whilst you tell yourself "I pick out not to overeat" the subconscious doesn't see the "not" it only sees overeat because it has a visible element. Words with visible problem appreciably impact the unconscious mind. With this in mind you'll want to be cautious with what you are pronouncing and anticipate.

2.The subconscious mind is aware of satisfactory NOW: Getting the unconscious thoughts to reap a intention set for the future might be a hint difficult, as it isn't always about the beyond or the destiny, it is simplest concerned with the prevailing. It is neither worried with beyond nor future activities. To gain dreams set for the future, you need to claim the very last results inside the gift. This is due to the truth the unconscious mind doesn't remember subjects as crucial until it is inside the gift. You will recognize that you are more privy to the outcome due to the fact the set-date methods. As your subconscious mind goes into motion it produces corresponding bodily reaction in shape of panic. The subconscious thoughts then begins offevolved to art work in a bid to ease the pain. This is why the catastrophe employee functions great beneath stress. Since the subconscious thoughts does no longer stay on the past, Therapies that revolve lots throughout the past might not be very effective.

three.The unconscious mind can't preserve invalidated beliefs: The unconscious mind is incapable of protective conflicting mind, once it accepts a perception; it handiest identifies with matters which may be in congruence with the perception and absolutely ignores a few thing that doesn't validate it. Until strong proof is supplied to dispute that belief, the unconscious thoughts will all of the time hold immediately to it. An instance of that could be a little one that has been informed that he's a very terrible infant and who hung on to that notion. The child's unconscious mind ought to most effective pick out research which might be in congruence with that perception and ignore opposite tales. Thus, the kid starts offevolved to show off dispositions of a horrific infant until that belief is invalidated. To invalidate awful mind and ideals one might need to become aware of evidence that disputes such records, like remembering the instances on the identical time as he has been accurate and kind to people. The 2nd there is powerful evidence to invalidate the notion, the perception is mechanically omitted.

4.The subconscious mind sees the maintenance of the idea of self as crucial: The subconscious mind will do everything it can to maintain the idea of self; this consists of the survival of the physical self. This is why it rejects thoughts contrary to its idea of self.

5.The subconscious mind sees CONCEPTUAL survival as emotional well-being or happiness: Your emotional properly-being is critical on your subconscious mind, so it'll obviously invalidate thoughts with the tendency of throwing you proper proper into a country of perpetual disappointment. And when you have stricken to look at, your subconscious doesn't deliver you sad mind except an internal stimulus or outdoor cause provokes the mind, however after some time your subconscious mind goals tries to make you feel better, but can also simply now not recognize why, this is why a few people that don't have any records of being suicidal need to devote suicide, it's far an offer from your subconscious mind to make you experience better.

6.The unconscious mind is ever inclined to offer a few detail outcomes in happiness and/or survival and will provide the impetus right away as quickly as it is acquainted with what is needed: This rule can be very much like rule 19 above. So, it's far your obligation to feed your unconscious thoughts with information with wholesome approaches to make you satisfied. You can log on to acquaint yourself with this, and as quickly as located say it to your self to convey it for your recognition. Once this is accomplished, you're in your manner to perpetual happiness.

7.The subconscious mind works simplest to your gain and does it 24/7: One of the topics if you want to need to make you observe via with Biohacking your subconscious mind is the reality that it's far constantly at artwork. So the immediate you learn to tap into the unconscious mind to get you some issue you need, you could get applicable information all day long, because the subconscious thoughts is the leader government of your dreamland, and sleep is, in fact, a more cushty mode for you, you will be relaxation confident of having some cool inspirations in your goals.

8.When it's miles aware about what you want to stay on and be glad, the unconscious mind presents that proper away: This rule reinforces guidelines 16 and 20. The moment your subconscious mind recognizes that you are sad or depressed, or that your survival is being compromised, it gets to art work proper now, and the immediate it has determined a way to make you revel in better, your recuperation begins right away. So sure, you can deliver it to your unconscious thoughts, guy is terrific set off.

nine.Once the subconscious thoughts updates a record to a right nation (happiness or survival), the file remains in the updated kingdom absolutely: This is one manner the unconscious mind maintains the concept of self and continues you strong. It continues updated documents i.E. Thoughts, mind, reminiscences in there completely and does now not discard except there is robust proof that it's far no longer capable of producing conceptual happiness or survival. Only then does the subconscious mind discard the antique thoughts and uploads the new report, producing a new sample of conduct.

Chapter 14: Biohacking your subconscious thoughts

Biohacking isn't a cutting-edge day concept, People have been at it for masses of years, experimented such quite a few techniques, tablets and techniques, there have been failures and successes, and nowadays we are capable of perceive some a success bio-hacks beginning from meditation and hypnosis to electric mind stimulation. Since the ones techniques have worked for a few human beings, it is able to be just right for you furthermore mght.

But in advance than you pass all out and dive into Biohacking your unconscious thoughts headlong, there are a couple of things you need to understand. The same manner earlier than hacking a tool, you must have vital information approximately how the tool works. The same goes for laptop systems, in view of the thoughts as a complicated pc, it's far vital that you examine a few subjects about it.

The Brain is the number one organ of the essential involved machine, the thoughts and the spinal cord makes up the important worrying machine. The Brain is made from cells known as neurons. Neurons are designed to transmit facts from the thoughts to different factors of the body and vice-versa to get numerous matters taking location inside the frame, much like the mail company. Communication from neuron-to neuron is made viable thru Neurotransmitters; the body's chemical messengers. Neurotransmitters convey all information and commands from the mind to every cell inside the frame. Biohacking the mind especially has to do with influencing the ones neurotransmitters. The predominant neurotransmitters consist of Acetylcholine, Norepinephrine, Serotonin,

Dopamine, Glutamate, Endorphin, GABA (Gamma-aminobutyric acid).

The identical way your thoughts is constructed to controls your bodily body, so is likewise the subconscious thoughts harassed to regulate physiological activities via a homeostatic impulse that keeps stability inside the frame, so that each one your body cells continuously function in a kingdom of harmony on every occasion. Your mind is constantly accepting and filtering statistics, to grow to be aware about the ones which is probably in alignment at the side of your pre-present beliefs. This is what office work your behavioral pattern and informs your idea of the concept of self. It is inside the realm of the unconscious that you may alternate your orientation and inform yourself of new strategies to gain happiness and all-spherical achievement. Here are five strategies wherein you can efficaciously bio-hack your Subconscious thoughts:

1.Meditation/ Yoga: Meditation is one effective way to bio-hack your unconscious thoughts. Studies have confirmed that other than internal peace, and the feeling of all-round wholeness that meditation lets in you gain, it moreover increases the grey matter variety density of the hippocampus this is accountable for mastering, retention and memory. Thus, permitting you to analyze new abilities without a problem, and consider a good buy of factors you study for a long time. Furthermore, meditation moreover allows to enhance awareness as it involves that specialize in a particular object for a long term.

Yoga is also effective in optimizing the subconscious thoughts functioning, as meditation and exercise are some of the components of yoga; wearing occasions help with the release of endorphins, and meditation permits to reinforce reminiscence, heighten focus and decorate hobby. Yoga has recovery results, daily practice of yoga has been determined to lessen growing older and reduce stress, and in reality, it grow to be determined that yoga reduces the signs of Post-demanding stress sickness (PTSD). PTSD is a illness that impacts humans who've suffered from horrible studies in the beyond that traumatized them; human beings living with this example normally have flashbacks and nightmares of the incident that brought approximately it. Hatha is a shape of yoga that lets in with PTSD.

2. Medication: One of the maximum well-known Biohacking system of the century are drugs, a few human beings opt for using high quality pills to other forms of treatment because of the truth they "truly

haven't were given the time". The use of medicine might be a quicker way to bio-hack your subconscious mind; it truely isn't always the maximum consistent. But if you decide to hack your thoughts using medicinal drugs, then you definately definately need to do so together with your clinical medical health practitioner's consent as some of those pills may be risky.

Two predominant schooling of drugs used for mind hacking are; Nootropics and psychedelics.

Nootropics are popularly called "clever capsules", and that's due to the truth they truly make you smarter. Nootropics artwork through developing mind efficiency. Some famous Nootropics embody: Modafinil, Forskolin, L-theanine, and Bacopa monnieri (water hyssop). Bacopa monnieri is a natural nootropic, it's miles greater stable to use because it has little or no thing effects except potential infertility this is said to be reversed after drug use is stopped. What's extra the effect of the drug stays even after discontinuation.

There are psychedelic drugs for thoughts hacks, those psychedelic drugs are typically applied in micro-doses. The motive for micro-dosing is for the affected man or woman to get first rate advantages of the drug without getting immoderate. Micro-dosing has received pretty a few floor in the Biohacking community as it has splendid consequences even as used to bio-hack the mind.

Some psychedelic capsules that may be used to bio-hack the subconscious mind encompass LSD AND Psilocybin. LSD is mainly used to decorate hobby; it helps you to interest higher. LSD dosage stages amongst 10-20 micrograms. Other benefits of LSD encompass better temper, expanded power and motivation, highbrow alertness.

Psilocybin is a psychedelic drug furthermore known as magic mushrooms. Micro-dosing psilocybin affects your cognitive capability drastically. Not first-rate does it assist you observe new topics with out problem, it additionally makes you feel happier and additional constructive, any other advantage of psilocybin is extended neurogenesis, it allows in the manufacturing of new neurons.

three.Hypnosis: Hypnosis is taken into consideration one of the fastest techniques to bio-hack the subconscious mind. Hypnosis permits you to achieve deep elements of your subconscious mind and get right of get admission to to long forgotten recollections, mind and feelings. Hypnosis is a method that works by way of setting you in a quiet and snug us of a, at the identical time because the voice of the hypnotist guides you rhythmically into the world of your mind. Hypnosis lets you get right of entry to the electricity of your unconscious mind and re-directs that power inside the course of attaining your set-dreams. Hypnosis permits to lessen pressure and tension, it moreover lets in patients laid low with diverse issues like sleep troubles and consuming troubles, hypnosis moreover can be instrumental to overcoming addictions. Hypnosis is the Biohacking tool you need to recover from traumas and generate extremely good strength that will help you collect your goals.

4.Diet: Your food regimen is pinnacle to Biohacking your mind, David Asprey, the pioneer of biohacking talked considerably approximately food that decorate thoughts interest. He came up with "The Bullet-proof Diet" a ketogenetic weight loss plan which includes avoiding sugar, gluten, maximum carbs and dairy, and consuming grass-fed meat and natural greens in large quantities. And, counter-intuitively, growing stages of fat from butter, ghee and coconut oil. Additionally, consuming meals wealthy in omega3 and extraordinary wholesome fats can keep dopamine levels active. Likewise, special elements containing magnesium, lithium, orotate and niacin will enhance thoughts feature.

5. Sleep: Mothers emphasize this, one too normally; early to mattress, early to upward push they say. So many human beings growth up and simply save you dozing, from overdue night time activities on campus, to taking home mountain masses of hard work, and at the surrender of the day we get little or no sleep, off to paintings and the cycle continues. Because mother isn't always there to stress you to

mattress doesn't mean it's far top sufficient to starve your self of accurate sleep. Doctors' recommendation 6 ½ to eight hours of sleep every day, so some thing you do make certain get no much less than 6.Five hours of sleep. Hold on a sec, that is a vicinity of a whole day! But how can one collect six hours of sleep while there's a lot to do. This is wherein right planning is to be had in, my colleague will say "prioritize your priorities", have a selected sleep and wake-up time, and try to keep to it. It is critical you get sufficient sleep every day because of the truth that is the time your thoughts cleanses itself, Neurons wash themselves clean of pollution and your body gets to lighten up. What's extra, snoozing strengthens your immune tool, making your body sturdy sufficient to combat infections which means you get to live longer.

Chapter 15: Effective Biohacking Techniques

Biohacking your subconscious thoughts includes re-programming your internal thoughts to do better; it includes maximizing your capability, attaining your height. However, optimizing the general standard performance of the mind isn't always an afternoon pastime, it requires dedication and resolves as our every day reviews strive to steer us otherwise.

The first step to getting access to and using the strength of the unconscious mind is to take away bad mind. Being pessimistic will pleasant make subjects worse due to the fact your body responds to thoughts for your unconscious sponsored through way of sturdy emotion. This method if you are consistent scared of being ugly, in any other case you feel ugly, over the years you will possibly surely begin looking ugly.

In this section, we're able to check twelve powerful Biohacking techniques:

1.The Countering Technique: The countering method is a very effective device in terms of getting rid of terrible mind. Whenever a awful concept comes to your mind, you counter it at once with a without a doubt excessive exquisite counter-concept. Your mind right away discards the awful concept for the motive that subconscious thoughts does not apprehend negatives. For instance, when you have an exam to write down, and you are already having bad thoughts like: "I am going to fail this exam", "oh, I am going to have a resit, regardless of the whole thing, I am no longer so clever", you counter such thoughts right now with exceptionally super thoughts, you pass "I am getting right away As in this examination, due to the reality I am so clever" or you could say " I am going to skip so nicely, I'll be granted a entire scholarship". You are going to need to be your very own hype-man, because of the fact particular topics take place at the same time as you expect proper thoughts.

2.The Delete Button Technique: The delete button method is a very powerful manner of countering terrible thoughts. As speedy as a terrible idea famous its way on your mind, you press the delete button right away. Picture your self smashing the concept to quantities, or you can don't forget your self burning it to ashes. Do this as regularly as you can consequently, purging your thoughts easy of negativity.

3.The Burning Desire Technique: Learn to typically again your mind with desire, due to the truth your unconscious thoughts works through acting on thoughts followed thru way of a burning choice. The very second you come to be passionate about your dreams, and your choice is to see it completed, your subconscious thoughts receives to art work and communicates all available alternatives and possibilities to the aware mind on a way to obtain the popular reason. People who emerge as the remarkable at what they do most effective emerge as the amazing because of their burning desire to attain their top. They get obsessed and grow to be so obsessed on their intention, they allow their wants to dominate their life and the whole thing else will become secondary, and that they emerge as willing to perform a little element it takes to understand their dreams. Cultivating a burning desire is straightforward, first you need to outline your intention and make sure it's far viable, be particular with what you need, Pick a pen and paper, then write it down, in the long run permit the desire to appearance it stand up consume you.

4.The Bridge Burning Technique: Do away with the safety boats you're preserving inside the event that a few component is going wrong, permit skip and burn your bridges. Don't be concerned which you'll crash down to earth due to the reality you received't. Our minds have been designed to help us stay on; it's far while you revel in you haven't any alternatives which to procure the brilliant consequences. Burning your bridges unlocks the survival mode, in which you could only flow into in advance toward your selected intention.

five.The Small wins Technique: The small wins method gets you to obtain your goals in a modern manner. Break your motive down into smaller viable desires to be completed within a completely unique time-body. It makes the entire approach appear like a chunk of cake. For example, when you have 5 books to examine in a month, looking at it from the attitude of

"Reading five complete books in four weeks" can also appearance no longer feasible, and you may end up no longer reading any. But at the same time as you smash it all of the manner right right down to one ebook in in line with week, and then further wreck it proper right down to a particular quantity of chapters an afternoon, it seems much less tough and additional ability.

6.The Motivational Technique: Human beings have the capability to do excellently well whilst they will be properly prompted. This technique makes you do loads more than you concept you are able to. This is why a few businesses present mouth-watering incentives to encourage their personnel. You also can appoint this technique to gather most productivity. Focus at the save you end result and the way adorable it's far, it'll propel you to do higher and additionally persevere till you benefit your purpose. Additionally, surround yourself with extraordinary reinforcement, this includes fantastic friends that could inspire you or a song or film that encourage you inside the route of project your goal.

7.The Visualization Technique: Look thru the eyes of your thoughts and do not forget your life after you've got achieved that dream. I inform you, the frenzy of emotion you sense is enough to propel you all of the manner thru the method. And to advantage this, you need to accept as authentic with that your desires will come real, not that it could come right, you want to guarantee your self and accept as actual with strongly to your capability to starting this dream. Because if you fail to consider in your self, whilst conflicting conditions come, you could get very discouraged and cease or worse, fail. So, take time-out each day to expect your self after your aim has been completed. Feel the right away in advance of time, agree with the Christian Louboutin red bottoms you can placed on at the crimson carpet, exercising the wave and smile, what you'll say to the crimson-carpet host. And in no time, it would become a truth.

eight.The Physical Preparation Technique: The physical instruction method sincerely requires you to make proper enough schooling inside the path of carrying out your dreams. What are the ones topics you may need in the course of and whilst you in the end finished your motive, write them down and get them. If you want to come to be a great model, as an example, begin working for your posture and your facial expressions, get prepared to be disciplined collectively with your food plan, get used to strolling out and maintaining a trim decide. Learn to be prepared for the entirety, so matters don't seize you unprepared.

9.The Detachment Technique: Firstly, there's a difference a few of the outcome of a choice and the actual fulfillment of your desires. Detach your self from results, however as an possibility interest for your dreams being fulfilled. Because, you cannot are expecting how precisely your dreams will come to expose up itself, topics exchange at each issue in time, so don't be too inflexible collectively along with your plans. Have an open thoughts and accept as proper with inside the popularity of your goals.

10.The Mantra Technique: This method is one of the most effective strategies utilized in Biohacking the unconscious mind. The method normally has to do with having you repeat a nice mantra for your mind. The essence of that is to position more faith for your phrases. The more you repeat it, the more you accept as proper with it, and the greater you bear in mind it the higher your chances of getting an brilliant end result. The mantra technique allows you conquer horrible emotion that get inside the way, emotions like fear, anger and sadness. Scientific research has shown that hypochondria is one of the primary reasons of ailments. Hypochondria is actually you convincing your self that you are unwell. It is a kind of car-idea mantra as a consequence of bad emotions like worry. Then if it's far feasible to influence your self which you are sick, it's far further feasible to convince yourself that you are sturdy and healthful.

eleven.The Reading out loud Technique: without a doubt because the mantra technique requires you to say remarkable mantras out loud however, for your thoughts, the studying out loud method requires you reading it out loud along with your mouth. Reading your goals out loud makes it much less complex for the aware mind to familiarize itself with them due to the reality your ear hears and relays the facts, this hastens right processing of information. So, after you have got were given set particular desires for yourself, make certain to install writing them down and decrease again them up with a burning preference to benefit it. When studying your desires out loud, be passionate and emotional about it. And lastly, be given as true with strongly that you will attain your set-desires.

12.The Celebration technique: Lastly, have a first-rate time your achievements regardless of how small. Celebrating your achievements makes you feel specific about your self and motivates you. Remember how happy you felt even as you have been little, and your mother and father may additionally come up with a piece of sweet at the identical time as you purchased your sums right? That is because of the fact every wonderful challenge deserves to be celebrated.

Chapter 16: What is Biohacking?

I do not apprehend if you are like me…. The first time I heard the time period "biohacking", I had visions of quick cuts for genetically engineering quieter puppies that don't shed. I had pictures of bespectacled mad scientists, slaving over Petri dishes yelling, "It's alive!" So I googled the time period, as you do, and the primary problem to come lower back up confirmed my mind: "biohacking" is described thru the Merriam Webster Online Dictionary as "the hobby of exploiting genetic material experimentally with out regard to not unusual ethical standards, or for criminal functions." So how have to this help me be a higher person, a happier, healthier man or woman?

Biohacking did virtually start inside the worldwide of genetics. For a number of us, genetic manipulation, wherein people in lab coats experiment with a topic's DNA to create the right baby or make a food crop proof against pests, is scary and first rate unexpectedly. As with any new discovery, it did not take long earlier than amateur biologists began wearing out their own experiments in basement labs and makeshift centers in their garages.

These beginner geneticists, the genuine "biohackers" (or "biopunks" as some are looking for advice from them) tinker with genetic code, DNA, proteins and micro organism. Only more than one a long time within the past, what they may do at home now changed into high-quality feasible in excessive-tech facilities. Thanks to less pricey, prepared-made kits available over the internet, it has end up viable with a view to clone and series a gene within the consolation of your own home in a whole lot much less than three days. It has even become alternatively an awful lot much less expensive—what used to charge $one hundred,000 (to collection DNA) now only expenses ten cents. Anyone with a biology degree (or maybe definitely properly-look at in the sciences) can end up a genetic detective. It is the traditional case of not needing a "why"— it's far because of the reality they'll be able to.

That form of biohacking stocks one problem in not unusual with the biohacking thoughts that you may use to enhance your existence. At its very center, it's far what you may practice to your very very very own biohacks—the choice to get to the inspiration of a few aspect extra quick, gathering statistics and the usage of the equipment you need to hand. Granted maybe genetic manipulation ought to make one taller or quicker or extra cute, but that software program program of the technological understanding is an extended manner off.

The listing of generation available to be had available on the market is significant. You can buy a system or an app for almost the whole thing because of the truth the market has exploded for some issue that allows you song your biodata. Using an app to track your sleep styles or a wrist watch that doubles as a coronary heart display are smooth and sensible techniques to take rate of your health. Not absolutely everyone can provide you with the money for them despite the fact that, or for that recall is attracted to that level of obligation.

No, the biohacking I am going to be teaching you approximately has to do with quick cuts to using what you already have to its remarkable advantage. I need to teach you approximately the usage of biohacks to meet the greater straight away, real existence pastimes of fitness, happiness and achievement.

Thanks to Mark Moschel in his put up for bulletproffexec.Com, we have the definition of this shape of biohacking as "the selection to be honestly the remarkable model of ourselves. The vital aspect that separates a biohacker from the rest of the self-improvement worldwide is a systems-questioning approach to our personal biology."

Still a piece unsure as to what it all way? Or how you can use biohacks to make your existence better? Do no longer be worried. It takes some work and locating the method that works incredible for you—in truth, it's miles all approximately you and your "quantified self". You can be examined the way to use meditation to combat ache if that appeals to you most; the manner to consume mindfully for your health; or a way to use song that will help you pass beyond your darkish mind.

No one seems to realize who coined the term "biohacker", but possibilities are it's far like a number of the contemporary "hacking" terms (i.E., the famous YouTube "existence hacks"). It have become stimulated by way of the use of the politically stimulated pc hackers of the last decade. Hacking is synonymous with locating quick cuts, decrease again doors and the fastest way to a effective aim.

Biohacking has superior out of the lab and into our everyday lives. It is an umbrella term, sincerely, one which covers masses of components of self-improvement. Today's speedy paced, driven society has created a populace of over worked, unstable, pressured out people decided to take control in their fitness. We need so one can live the richest, healthiest lifestyles we are able to.

Enter biohacking. As I use it in the course of the e book, it covers all the techniques wherein you may enhance your probabilities at success. Things like taking a yoga splendor to alleviate your aches and pains or adopting a vegan weight-reduction plan that will help you drop the ones closing five kilos and offer you with glowing pores and pores and pores and skin are all perfect options. It calls for that you get in touch together with your thoughts-body connection, to sincerely reach into your self and turn out to be aware of strategies your frame and mind want to paintings in concord if you want to be the splendid "you".

Anyone can meditate, skip Paleo or start Pilates. But to get quicker effects, you may need to use a biohack. And the subsequent chapters will let you do really that. I am going to provide you several tips and techniques that, in case you are inspired, can change your life for the better.

The Quantified Self

The size of personal information and outcomes has never been tons less hard. Now everybody may additionally want to have access to pretty specific, problematic stages of records. In fact, this is wherein technology comes into play (the relationship to the era of biohacking). The explosion in wearable devices is a extraordinary example of this. Forget the primary such invention—the calculator watch of the Eighties—now you have were given clever apparel, wearable coronary coronary heart price video show devices, and step counter, and many others.

The most updated wearable tech has made technology pervasive in normal life. Whether it's a piece of secret agent tech or a temperature-sensitive track in form, it has in no way been an awful lot less difficult to quantify yourself, your very personal character data if you'll. "Self-information thru self-monitoring" as Mark Moschel so eloquently framed it in his submit "The Beginner's Guide to Quantified Self" (blog.Underarmour.Com). The quantified self has clearly become a movement of kinds. At its middle is the usage of generation to collect facts on oneself. The facts can be done to everything from weight reduction to struggling with continual pain to becoming greater a success. Being capable of acquire information on yourself permits you to "quantify" your fitness greater precisely.

You name it, you could music it—your hydration stages, sleep patterns, your caloric intake/output, and so forth. It is like collecting ammunition for a few aspect you'll be stopping. Granted you can end up crushed if you try to cope with all problems right away. Tracking may moreover need to in reality add for your stresses and now not every body may have enough cash the present day-day tech. That is why I inspire a not unusual sense approach to biohacking. It is as plenty as you to pick out what it is you're looking for earlier than you start biohacking it. And that consequences in the number one detail of biohacking: mindfulness.

Biohacking works extremely good if you have a certain quantity of self-focus. You will need to start making connections amongst movements and outcomes. For example, making the connection among what you placed into your body and the response your frame has to it. Biohacking works superb with forethought and planning.

That said, biohacking appears to have taken the best foothold in the fitness business enterprise. I desire to provide you with numerous hacks you can out into exercising right away. The concept of preserving music of your personal body's statistics isn't new—it's miles stated that even Benjamin Franklin saved a log of wonderful non-public virtues to resource him in his quest for ethical perfection. Technology has simply driven us that a good deal further and deeper into the understanding of ourselves, our biometrics.

You do not need to hurry out and buy a Fitbit although. If Benjamin Franklin may additionally additionally want to maintain tune of his facts with a quill and a magazine, you may really tune results on an iPad or maybe in your private mag. You might also moreover already be tracking certain physical states: possibly you have to show your blood stress in the course of the day or your blood sugar in advance than each meal. If you aren't inside the addiction, I do endorse you to discover a way of quantifying and recording a few element it is that you are trying to attain (weight reduction would possibly entail quantifying caloric intake, exercising or energy burned and so on). Biohacking suggests you a manner to use data to make the brilliant picks in your health, mind and fulfillment. You will learn how to use experimental processes to perceive what works amazing for you.

Chapter 17: Biohacking in Action

Everyone seems to be seeking out something these days—the ideal way to those weight, the quickest path to happiness and the appropriate manner to turn out to be a hit. In the following few pages, I will deal with some of troubles that you may need assist with and the way biohacking can useful resource for your achievement.

Chronic Pain

Chronic pain is at epidemic tiers in North America. From migraines to arthritis, the growing older population and the compelled out body of workers are all suffering. Besides making clean physical responsibilities tough or not possible, pain causes highbrow fogginess. I in my view describe my persistent ache as managing a normal humming in my head. It makes it hard to pay interest because of the reality your mind can not suppose through the aches and pains. We all recognize how tough present day-day day human beings are locating it to live centered—at work, at school or maybe at domestic. We find our minds racing round the whole lot however the second accessible.

So how are we able to use biohacks to conquer persistent pain? One of the first-class matters that has come from my struggles is extra suitable frame attention. Remember the "quantified self"? In order to conquer a few difficulty, biohackers first recognize and quantify the hassle. I were compelled to endure in mind and experience into the internal most muscle agencies, imagining the nerve endings sending flares into the encompassing joint, inflicting pain and burning in their wake. This recognition has sincerely led me as a way to fight the pain with my mind.

A proper location to begin is along with your posture. Long long past are the times at the same time as posture have become one of the many suitable behavior taught in university. Sadly, for generations now we have were given gotten regularly slouchier. Our tech obsessions have carried out the bulk of the harm within the last two a long time, with that first gaming console and desk pinnacle computer to the smartphones and tablets.

Why is posture important? The crucial motive is your spine is much like the trunk of a tree—it permits the relaxation of the limbs. And it is a primary nerve pathway. If it's miles broken or out of alignment, the ensuing pain is like dominoes—it cascades into the relaxation of your frame.

Changing topics to keep away from lousy posture is less complex said than achieved and also you might imagine it not possible in instances which includes running in a sedentary assignment. One biohack you could use is an exercise to stabilize your spine. In a seated role, clench your buttocks, then take a deep breath thru your nose, pulling your ribcage in and up. Exhale and allow the ribcage drop down. Keep your belly muscular tissues engaged. Then align your head in what's called a "impartial function", which means your ears are consistent with your shoulders. Do this workout whilst you first sit down and attempt to preserve your belly muscle businesses twenty percentage engaged at a few stage within the day. Stabilizing your backbone even as you first sit down down makes you aware about your posture and consequently lets in you to adjust it while it slips. You can test-in with your self each hour or so and modify as wished.

Another biohack to make certain a wholesome spine includes a pool noodle. I located this one from my physiotherapist. Stand along with your lower back in the direction of a wall. Place a pool noodle among the small of your lower returned and the wall and press toward it to maintain it in vicinity. Keep your toes shoulder-width apart. Align your ears over your shoulders and take a deep breath as informed above. Slowly slide your self down the wall, allowing the pool noodle to gently rub down your spine. Come to a squatting position after which slowly flow decrease lower back as plenty as standing. Do this as quickly as an afternoon for every week and you will be amazed how this biohack enables with again ache.

Other styles of continual ache requires the identical mindfulness to combat. Track the times of day and the movements that would have introduced on a flare up. In doing so, you may possibly have a observe a sample. Once you be aware the pattern you can most likely recognize the reasons and therefore is probably able to discover decision. You may additionally moreover furthermore want to transport see your scientific medical doctor if clean wearing occasions do not artwork, but you'll be capable of extra sincerely describe your condition because of the records you can have accumulated.

Practicing any form of rest techniques are also appropriate hacks for tolerating pain. Ensure you in no way push too some distance even though. Sometimes in our eagerness to have effects, we are able to damage ourselves. Take it slow and listen to your frame. Meditation along physio is a mainly effective exercising (see below for hacks for meditating).

Once you've got the pain under manage, deciding on up a step counter might be a fantastic idea to maintain track of the manner lively you are in a day. The records accrued from that is probably used to inspire a modern-day exercise routine. Always be seeking out approaches to art work a similarly step into your day—taking walks is one of the remarkable physical video games there can be.

Biohack Your Meals and Nutrition

Are you worn-out all the time? Plagued through complications? Tackling embarrassing and painful digestive issues? Most oldsters in North America are combatting one or all of these troubles. So what are you able to do to get short relief and lasting effects?

Biohack your weight loss plan. I am no longer announcing flow into on a eating regimen however observe your gift ingesting conduct (and patterns). Too numerous us consume with out a belief to what we are setting into our our our our bodies. We are so busy we sometimes bypass food in truth. We depend upon processed components an excessive amount of. So what are you able to do for yourself to get a tremendous start on ingesting better? Axe the sugar. By now you can not stay in North America at the same time as no longer having found out that too much sugar is a killer. But have you ever ever ever in reality absorbed the harm your midday pick out-me-up chocolate bar and soda is doing, particularly long term? Simply positioned, we consume too much sugar, intentional and hidden, and we want to save you.

Removing sugar from your eating regimen isn't always difficult, if you teach yourself about the food you placed for your dinner table or on your lunch subject. It is extra the withdrawal that motives the problems and ultimately can smash your risk at success. That is why it's far important to avoid the rebound complications and tiredness thru going slowly. Biohacking your diet regime begins with monitoring the sugars you are eating (or consuming).

You should take every week and record the entirety you consume, at the side of the caloric content material of the gadgets, paying precise hobby to the sugars of course. Then take according to week and decrease out the extra sugar you use—choose out out of one of the sugars you typically established your morning espresso or change your midafternoon snack from a chocolate bar to a bit of low sugar fruit or a few raw greens. Try this easy change for constant with week then boom the amount you get rid of. You're your coffee black or pick out juice as opposed to soda. Keep progressing and noting the manner you experience— make sure you're making the relationship that what you are not installing your body is making you feel greater wholesome. Reinforce the choice to achieve success.

Eliminating sugar is a amazing the first step. Next you need to biohack your diet everyday. The outstanding hack I realize for this is, all over again, pretty easy: cast off processed food from your refrigerator and freezer. As I sincerely have stated, take stock of the nutritional labels to appearance in reality what rate the ingredients have. In their area load up on generous helpings of colourful vegetables and fruit. Another proper pass is to replace margarine with actual butter. The hack works because you're giving your body the gasoline it may use and needs whilst you supply it more herbal vitamins.

Of direction, in case you are a meat eater, update the ones premade hamburger patties with domestic made grass fed beef patties. Add low mercury fish (like salmon) and unfastened range hen to your meal plan. Be privy to the quantity of starches you eat (rice, potatoes and so on). Pay interest to portions and try to consume first-rate long grain variations of rice.

Remember, the purpose is to devour as close to natural as possible. And due to the truth you still want sugar to your weight loss program however do not want the processed kind, get your energy from low sugar end result (strawberries and lots of others).

Biohack with the resource of going certified herbal if you could. Look for grass fed red meat, unfastened range eggs, and many others. If that isn't feasible then genuinely attempt to buy meals which is probably as near their precise u . S . As viable. For example, frozen veggies are a extraordinary alternative for clean and are a better substitute than canned. Frozen fruit is likewise an notable preference.

Another dietary hack that focuses on your individual desires is making an attempt out your self for meals sensitivities. Food sensitivities can reason a massive amount of issues, from fuel and bloating to rashes and complications. Now no longer something replaces the care of a scientific doctor, however you can begin the way of checking out at home. Instead of subjecting your self to reams of checks right away, you may begin a preliminary assessment of your very very very own. It is as easy as removing the most effective meals which you suppose may be at the inspiration of your digestive upsets for each week.

Biohacking comes into play because of the truth you will be attentive for your body. Let's say you pick to take away dairy out of your diet due to the truth you have got completed your research and suspect it could be the cause you are having belly pains.

Chapter 18: Get Back to Nature

The past thirty or so years have visible North Americans come to be greater domesticated than ever in advance than. We have grow to be a society afraid of the outside. We have made our lives sterile and the result is a unwell population. We wash the whole thing with anti-bacterial cleaning soap, we spray insecticides on our lawns due to the fact we've true advantageous flora as "weeds" and we nearly bubble wrap our youngsters for fear of dust and bacteria. The quit end result: we're making ourselves sick. We are lots much less healthy, a bargain much much less resilient and lots lots less adaptable. More humans are stricken by continual conditions than ever earlier than.

So biohackers like Daniel Vitalis, a well-known health motivator and strategist, have determined we need to "re-wild". Much like the hippie movement of the

'60s and '70s, we want to head again to nature if you need to discover balance in our lives. Take my advice and try the subsequent 4 smooth hacks, so that you can pass returned to nature and because of this regain that resilience of previous generations:

1.Eat tons a lot much less processed meals. Forage for wild, sparkling elements if you may. If you advise to strive foraging, make certain you have got completed your research or higher however discover an expert within the safe to consume wild food to be had in your vicinity. You do not want to get ill from consuming the incorrect plant. Not truly every person has inexperienced spaces that allow for this pursuit however you could still add extra raw vegetables and give up result to your food regimen to have some of the blessings.

2.Drink unprocessed water. Tap water is so intently handled that maximum of the minerals have been stripped away, so a better possibility is spring water. This might not be an possibility because of economic problems or perhaps ecological issues approximately the shape of bottles going into landfills, but if at all possible, you need to attempt to get some shape of herbal, mineralized water into your weight loss program.

3.Get smooth air. Get outside. The air in our offices and homes is stagnant and regularly recirculating bacteria and airborne debris, for this reason contributing to the growth in respiration ailments. If the least bit possible, make your outdoor journey inside the u . S ., far from the smog and pollutants of the town streets.

four.Let the mild in. Human beings can not make their non-public Vitamin D, as a quit result they need daylight hours to live healthy. Vitamin D is mainly important for maintaining melancholy at bay. So get outdoor for as a minimum ten minutes a day to your Vitamin D recuperation. And every different tip is to now not typically placed on your sun shades. Much of the Vitamin D is absorbed thru our eyes so permit them to be naked for a hint every day. If that isn't always possible, communicate to your medical doctor approximately getting a sunlamp.

If you check, all of those hints include getting outdoor. We want to get out of our concrete jungles and into the geographical location more. Again, we're often constrained because of our locale. So in case you cannot get out of the metropolis, strive finding a park in the metropolis or possibly a botanical garden to biohack your want for nature. A very last observe: do now not be terrified of dirt or of your kids gambling within the dust. It is now being identified that a touch publicity is going an prolonged manner to constructing sturdy immunities. And stay some distance from anti-bacterial the whole lot—some of the wipes you buy in reality contain insecticides and no individual goals greater exposure to chemical substances.

Mental Health Biohacks

So you have were given sorted your body and your vitamins, but you still do no longer enjoy well. Life however sucks. Well, it may be time to biohack your attitude: change your body of reference and your expectations.

Thousands of authors have examined what it is that makes us "glad". And of direction, the way to get glad. You can strive each short restore and observe every one of the seven steps to happiness, but ultimately if you do no longer alternate your mind-set, no longer something will art work. It is like putting a band beneficial resource on the problem. If you do no longer discover and heal the trouble, it'll come once more.

Biohacking guru Joe De Sena, creator of the Spartan Races, has found we want to be extra like generations before us. It sincerely appears to be a imperative them to the biohacking world. Our grandparents got masses right at the same time as it came to dwelling satisfied and healthy. We, however, have emerge as a society of excessive expectations for little effort. We can tolerate very little inside the manner of stressful conditions or road blocks. Our pores and pores and skin isn't always as thick as our parents'.

If you need to biohack your temper and mind-set, you need to become privy to your mindset. How do you commonly cope with disturbing situations or sudden occurrences? Do you spin out of manipulate, adopt a fatalistic attitude or do you notice the possibilities for decision? Are you a pitcher half of of-empty or glass half of of of-whole sort of man or woman? If you are searching out assist, you then definately are most probably a tumbler 1/2 of-empty type of man or woman. You are not on my own. Too plenty of us find out ourselves looking at the arena with a enjoy of hopelessness. We discover even the slightest bump in the street throws us off target. We get stuck in an endless cycle of fatalism, thinking negatively. But you may exchange the report.

The first step, recall it or not, is to intentionally make your self uncomfortable. This is a completely personal pursuit, however a few easy hacks could be to expose your self to a few issue you locate ugly. So take a chilly shower, run a 1/2 of marathon, recollect your personal death, and so forth. You get the photo. How does this art work? Well, it's far like aversion therapy.

Aversion treatment is a behavioral alternate technique used by therapists to overcome addictions, obsessions and even violent behaviors. When it includes biohacking, you may use it to overcome fears and alternate your mind-set approximately your existence in wellknown. The same principle is at art work: even as you face your fears and live on, you need to now not worry them.

You'll see that you could tolerate the stuff you fear however you can moreover be better equipped to comprehend what you do have. In a weird way, this effects in another biohack—meditation. Remember I said mindfulness is a key to powerful biohacking? Meditation is the very essence of that. The act itself is a lot tons less complex than you may assume. In a awesome scenario, you may be able to take a seat effects in a room via yourself, with calming tune and incense burning. But you may carve out time to meditate irrespective of how busy your day is probably. Hack it.

I need to take several 5 minute periods within the route of my day to certainly close to my eyes and song out to everything. Then I silently repeat "thank you God" over and over, permitting myself to pay attention most effective on the terms and enjoyable my muscle companies from head to toe. Learning isolation techniques allows—wherein you consciousness on factors of your body separately (typically out of your toes to your head). You demanding for a few seconds then release.

Chapter 19: Biohack Your Brain

Meditation is also extraordinary for enhancing recognition. And it is one of the stronger techniques to awaken sensory recollections—anyone has that one tune that takes us once more to a certain point in time and so forth. But did that during accord with song it's far that much more effective? Biohacking with music honestly stems from medical experiments which have proved track can exchange your mind waves.

Think of your thoughts as a community of electrical cables firing on the rate of milliseconds. In fact, your thoughts has 100 billion neurons the use of electricity to "communicate" to each distinctive. The dimension of that hobby with an EEG (electroencephalogram) shows the pattern as a thoughts wave.

What you're doing or wondering at any individual time dictates that mind wave hobby. Simplistically, the ones mind waves are labeled in four states:

1.Beta: that is the dominion you are in at the same time as you are alert, wide awake and concentrated on a assignment. This is your "regular" nation (14-30Hz).

2.Alpha: this may be the prevent of day country—you are although wide awake but extra cushty. Often this can be in that grey place between waking and snoozing. Memories regularly come flooding in while you are in this kingdom (eight-14 Hz).

three.Theta: this is your subconscious thoughts at art work, at the same time as you might experience slight dreaming (four-eight Hz).

four.Delta: that is deep sleep. A dreamless us of a, this is while your brain and frame get well from the day's difficult work (zero.1-4 Hz).

In order to biohack your way into a particular united states of america of the us of thoughts, you can use the era of tune. Using tune that fits the mind wave sample (Hertz) can inspire a specific country of thoughts. A lot of work has been finished with binaural beats in association with meditation, in fact.

A binaural beat is composed basically of one in every of a kind frequencies, every one emitting from a different headphone honestly so they may be separated. Your mind then creates a contemporary frequency from the 2 (it truly is the distinction the various ones preliminary frequencies). For instance, a frequency of 205 Hz that comes via the left channel and a frequency of two hundred Hz that comes through the right channel could make the mind create a frequency of 5 Hz (205-2 hundred=5).

By doing this, we are capable of create infrasounds, reputedly imperceptible sounds to our ears, but our brains sign up them. The mind makes use of infrasound frequencies itself relying on in which state it's far in (awake, snug, sleep, and dream). Through inducing amazing frequencies artificially you can bring your self proper into a rustic of relaxation or maybe a dream-like state and because of this alter your recognition and cutting-edge perception. When applied in coordination with meditation, song can decorate the depth of rest you experience. You can find all precise varieties of those beats at the net.

One factor that I surely have not but stated however that ties in with the concept of manipulating your mind waves is sleep. Or lack thereof almost about most folks these days. This biohack using tune to growth your meditative country additionally can be done to growing the proper surroundings for an brilliant night time time's sleep. You can hire beats to inspire your mind to settle and for that reason get a deeper, greater recuperative sleep. This might also in reality be the best recovery in case you are on shift artwork. Because our our our bodies aren't properly-appropriate to snoozing in a few unspecified time in the future of the day (because of our circadian rhythms), you may use this hack to ensure you continue to get relaxation even though it is sunny outside.

Biohack Your "Flow"

One of the modern topics to hit the "locating happiness" circuit is a few trouble called "waft". It also can sound a piece indistinct or perhaps romantic however it's miles really quite plenty that feeling you get at the same time as you lose yourself in an hobby. For some people, developing track allows them to go along with the glide. For others it might be walking on a traditional car. No rely the movement, there is usually a motive, a project preceding to locating that kingdom. There is the feeling of needing to conquer some thing that allows you to be happy. It is the overcoming of that venture that allows you to discover your flow in the decision. It is a cycle of types.

The simplest manner to biohack your float is to push your self into a undertaking— whilst your brain is provided with a trouble, it enters the reading phase (beta). Pushing into it, accepting the disappointment, allows the mind to shift from studying to unconscious (alpha). This is in which the magical america of drift in reality takes vicinity. In alpha, your thoughts is snug and therefore open to running at the statistics. Theta waves (subconscious) take over in the end, once the undertaking is completed. Then your thoughts enters the healing segment (delta).

You can biohack your waft while given a typically daunting mission by manner of pushing yourself further and longer than you usually might. Test your preconceived barriers. Then switch gears for a time—go away the challenge and visit a few issue mindless, allowing your subconscious to paintings at the problem in the historical beyond, so to speak. I locate if I am ever furnished with creator's block that if I bypass carry out a hint laundry it seems to loose me from the block. I am capable of come lower again to the assignment with new mind and electricity. Maybe for you this would be an notable time to get that sunshine? Once you discover that state of rest, your mind will will let you comprehend while it is ready to move lower back to the assignment.

Biohack Your Happiness

If you circulate lower back over all I actually have written to date, you may see the relevant problem rely for my biohacking pursuit centers in the long run on happiness. The very last biohack to make sure happiness itself centers on gratitude. To my mind gratitude is some component I was satisfactory capable of feel or display while matters have been going proper. I changed into now not a hopeful individual and even as matters had been given difficult, I felt that the whole thing have come to be hard. When the chips had been down, I positioned it tough to be grateful. And I am amazing I am not on my own.

That is why biohacking your manner to gratitude may be the exquisite issue you can do for your self. And it is as easy as switching gears or converting the channel. Instead of framing your reviews inside the terrible, exchange the wording to the extraordinary. That is the hack.

For instance, say you're going for walks overdue for paintings and the barista at your selected espresso hold is a trainee… sufficient said. He has tousled each different purchaser's order and is having to remake the drinks. What is your default reaction to this scenario? Did you robotically annoying up and clench your tooth? Did you heart fee upward push a hint taking into consideration the day this occurred to you? Did you trust you studied horrible thoughts?

Then trade that perception pattern from "did they should placed a beginner on in the course of rush hour" to "applicable aspect they have got employed greater body of workers to help them in the mornings—as quickly as they will be knowledgeable subjects will pass extra with out issues." Or take this time to be high-quality—do a brief meditation. You want to select to be extraordinary. You can find the good in most matters.

Gratitude also can take some time earlier than it becomes automatic, especially if you have spent a wonderful deal of your person life reacting to surprising conditions. You can retrain, even though. Some beneficial physical sports activities encompass:

Keeping a gratitude magazine. You do now not have to write prolonged, wordy entries. Just commit to writing down 3 belongings you are thankful for every day. As with many stuff, exercise makes perfect. When you are having a lousy day remind yourself of the gives you do have.

Take a each day gratitude stroll. You'll be hacking as a minimum 3 matters in case you do this (the essence of biohacking). The timing is as a lot as you, however you may benefit most at the cease of your art work day. The excellent region might be to get outdoor so you need to take inside the herbal environment, drink within the colorations and smells round you. If getting out of doors is difficult, attempt to at least take your walk in a selected setting—numerous department stores offer strolling programs while they may be closed. Even on an indoor walk, you may be capable of exercising your senses. Tune into your body as you stroll, center yourself and deal with your respiration. Be thankful deep interior. Say a mantra of gratitude.

Write a gratitude letter. This isn't a "thank you" word like you could write after a person gave you a birthday gift nor do you want to try this on a every day basis. But it's far an specifically powerful gratitude exercising. I undertaking you to undergo in thoughts a person for your existence that you are feeling grateful for. Not everyone is lucky sufficient to have a person right now, however it could also be a person who handed away. Write them a letter outlining what it's far they gave to you and your life that makes you grateful for them. If you are fortunate sufficient to have someone in your lifestyles right now, you can take this exercise to a few other diploma via means of creating a date for really the 2 of you to satisfy and you can have a look at the letter aloud to them. It will assist you experience the depth of your gratitude and as a bonus you'll probable pass that gratitude at once to them. They may be stimulated to put in writing down down their private gratitude letter.

www.ingramcontent.com/pod-product-compliance
Lightning Source LLC
Chambersburg PA
CBHW062139020426
42335CB00013B/1258